KU-575-447

Collins

GCSE 9-1
Combined Science
in two weeks

Tom Adams
Dan Evans
Dan Foulder

Revision Planner

BIOLOGY

Note: You will see the following logos throughout this book:

HT – indicates content that is Higher Tier only.

WS – indicates 'Working Scientifically' content, which covers practical skills and data-related concepts.

Revision Planner

PHYSICS

Structure of cells

Prokaryotes

Prokaryotes are simple cells such as bacteria and archaebacteria. Archaebacteria are ancient bacteria with different types of cell wall and genetic code to other bacteria. They include microorganisms that can use methane or sulfur as a means of nutrition (methanogens and thermoacidophiles).

Prokaryotic cells:

- are smaller than eukaryotic cells
- have cytoplasm, a cell membrane and a cell wall
- have genetic material in the form of a DNA loop, together with rings of DNA called plasmids
- do not have a nucleus.

The chart below shows the relative sizes of prokaryotic and eukaryotic cells, together with the magnification range of different viewing devices.

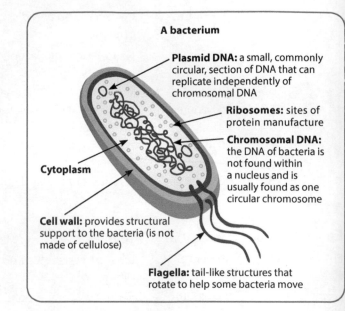

A bacterium

Plasmid DNA: a small, commonly circular, section of DNA that can replicate independently of chromosomal DNA

Ribosomes: sites of protein manufacture

Chromosomal DNA: the DNA of bacteria is not found within a nucleus and is usually found as one circular chromosome

Cytoplasm

Cell wall: provides structural support to the bacteria (is not made of cellulose)

Flagella: tail-like structures that rotate to help some bacteria move

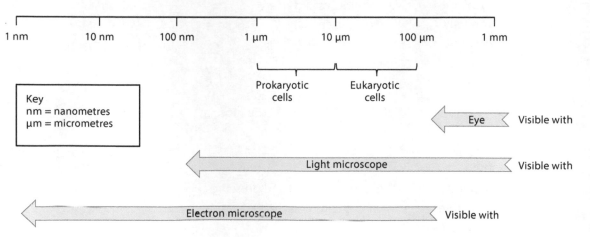

1 nm   10 nm   100 nm   1 µm   10 µm   100 µm   1 mm

Prokaryotic cells   Eukaryotic cells

Key
nm = nanometres
µm = micrometres

Eye — Visible with

Light microscope — Visible with

Electron microscope — Visible with

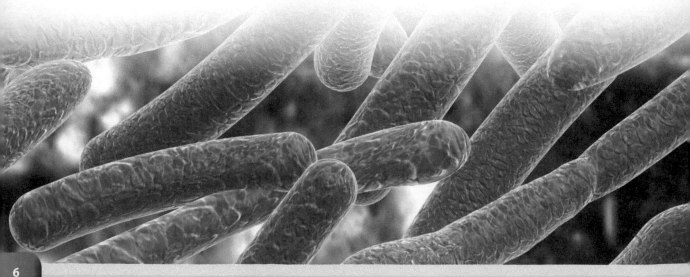

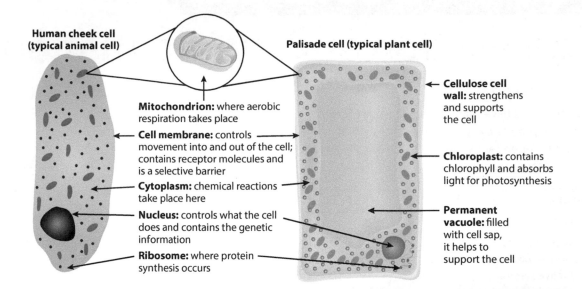

Human cheek cell (typical animal cell)

Palisade cell (typical plant cell)

Mitochondrion: where aerobic respiration takes place

Cell membrane: controls movement into and out of the cell; contains receptor molecules and is a selective barrier

Cytoplasm: chemical reactions take place here

Nucleus: controls what the cell does and contains the genetic information

Ribosome: where protein synthesis occurs

Cellulose cell wall: strengthens and supports the cell

Chloroplast: contains chlorophyll and absorbs light for photosynthesis

Permanent vacuole: filled with cell sap, it helps to support the cell

Eukaryotes

Eukaryotic cells:

- are more complex than prokaryotic cells (they have a cell membrane, cytoplasm and genetic material enclosed in a nucleus)
- are found in animals, plants, fungi (e.g. toadstools, yeasts, moulds) and protists (e.g. amoeba)
- contain membrane-bound structures called **organelles**, where specific functions are carried out.

Pictured above are the main organelles in plant and animal cells.

Plant cells tend to be more regular in shape than animal cells. They have additional structures: cell wall, sap vacuole and sometimes chloroplasts.

QUESTIONS

QUICK TEST

1. Prokaryotic cells are more complex than eukaryotic cells. True or false?

2. All cells have a nucleus. True or false?

3. Which organelle carries out the function of protein synthesis?

EXAM PRACTICE

1. a) Bacteria contain circular sections of DNA within their cells.
      Name these structures. **[1 mark]**

   b) Plant cells differ to animal cells in which one of the following ways? Tick **one** box. **[1 mark]**

      They contain a nucleus. ☐

      They do not have a cell membrane. ☐

      They possess a cell wall. ☐

      They contain ribosomes. ☐

   c) Muscle cells contain many mitochondria.
      Explain why this is. **[2 marks]**

   d) The average magnified length of a bacterium is 10mm.

      Using a magnification of 500, calculate its **actual** length in **μm**
      (1 μm is 1×10^3 mm).
      Show your working. **[2 marks]**

SUMMARY

- **Cells are the basis of life. All processes of life take place within them.**

- **There are many different kinds of specialised cell but they all have several common features.**

- **The two main categories of cells are prokaryotic (prokaryotes) and eukaryotic (eukaryotes).**

Organisation and differentiation

Cell specialisation

Animals and plants have many different types of cells. Each cell is adapted to carry out a specific function. Some cells can act independently, e.g. white blood cells, but most operate together as tissues.

Type of specialised animal cell

Sperm cells

- They are adapted for swimming in the female reproductive system – mitochondria in the neck release energy for swimming.
- They are adapted for carrying out fertilisation with an egg cell – the acrosome contains enzymes for digestion of the ovum's outer protective cells at fertilisation.

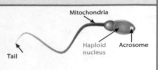

Egg cells (ova)

- They are very large in order to carry food reserves for the developing embryo.
- After fertilisation, the cell membrane changes and locks out other sperm.

Ciliated epithelial cells

- They line the respiratory passages and help protect the lungs against dust and microorganisms.

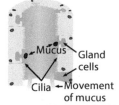

Nerve cells

- They have long, slender extensions called axons that carry nerve impulses.

Muscle cells

- They are able to contract (shorten) to bring about the movement of limbs.

Type of specialised plant cell

Root hair cells

- They have tiny, hair-like extensions. These increase the surface area of roots to help with the absorption of water and minerals.

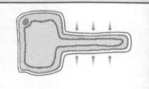

Xylem

- They are long, thin, hollow cells. Their shape helps with the transport of water through the stem, roots and leaves.

Phloem

- They are long, thin cells with pores in the end walls. Their structure helps the cell sap move from one phloem cell to the next.

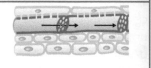

Principles of organisation

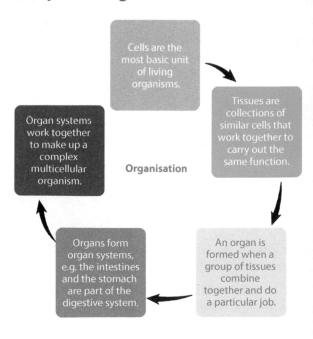

Cells are the most basic unit of living organisms.

Tissues are collections of similar cells that work together to carry out the same function.

Organisation

An organ is formed when a group of tissues combine together and do a particular job.

Organs form organ systems, e.g. the intestines and the stomach are part of the digestive system.

Organ systems work together to make up a complex multicellular organism.

Cell differentiation and stem cells

Stem cells are found in animals and plants. They are unspecialised or **undifferentiated**, which means that they have the potential to become almost any kind of cell. Once the cell is fully specialised, it will possess sub-cellular organelles specific to the function of that cell. Embryonic stem cells, compared with those found in adults' bone marrow, are more flexible in terms of what they can become.

Animal cells are mainly restricted to repair and replacement in later life. Plants retain their ability to **differentiate** (specialise) throughout life.

Using stem cells in humans

Therapeutic cloning treats conditions such as diabetes. Healthy pancreas cells can be cloned in order to achieve this. Embryonic stem cells (that can specialise into any type of cell) are produced with the same genes as the patient. If these are introduced into the body, they are not usually rejected.

Treating paralysis is possible using stem cells that are capable of differentiating into new nerve cells. The brain uses these new cells to transmit nervous impulses to the patient's muscles.

There are benefits and objections to using stem cells.

Benefits	Risks and objections
● Stem cells left over from in vitro fertilisation (IVF) treatment (that would otherwise be destroyed) can be used to treat serious conditions.	● Some people believe that an embryo at any age is a human being and so should not be used to grow cells or be experimented on.
● Stem cells are useful in studying how cell division goes wrong, e.g. cancer.	● One potential risk of using stem cells is transferring viral infections.
● In the future, stem cells could be used to grow new organs for transplantation.	● If stem cells are used in an operation, they might act as a reservoir of cancer cells that spread to other parts of the body.

Using stem cells in plants

Plants have regions of rapid cell division called **meristems**. These growth regions contain stem cells that can be used to produce clones cheaply and quickly.

Meristems can be used for:
● growing and preserving rare varieties to protect them from extinction
● producing large numbers of disease-resistant crop plants.

SUMMARY

● **Multicellular organisms need to have a coordinated system of structures so they can carry out vital processes, e.g. respiration, excretion, nutrition, etc.**

● **Differentiation is the process whereby cells become specialised.**

● **Stem cells have the potential to become almost any kind of cell.**

● **There are benefits, risks and objections to the use of stem cells.**

QUESTIONS

QUICK TEST

1. Which type of specialised animal cell helps protect the lungs against dust and microorganisms?

2. Which type of specialised plant cell is used to transport water through the stem, roots and leaves?

3. Name the regions of rapid cell division in plants.

EXAM PRACTICE

1. Name the process by which cells become specialised to carry out a particular function. **[1 mark]**

2. A fertilised egg cell is called a zygote. It is a type of stem cell.

   a) The zygote will divide and then produce specialised cells. Some of these cells will become neurones. Describe **one** adaptation of a neurone. **[1 mark]**

   b) Describe **two** applications of stem cell research. **[2 marks]**

   c) Explain why some people have objections to stem cell research. **[1 mark]**

Microscopy and microorganisms

Microscopes

Microscopes:

- observe objects that are too small to see with the naked eye
- are useful for showing detail at cellular and sub-cellular level.

There are two main types of microscope: the **light microscope** and the **electron microscope**.

Light microscope

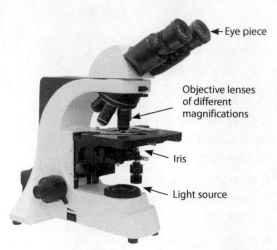

- Eye piece
- Objective lenses of different magnifications
- Iris
- Light source

Electron microscope

You will probably use a light microscope in your school laboratory.

The electron microscope was invented in1931. It has increased our understanding of sub-cellular structures because it has much higher magnifications and resolution than a light microscope.

These are white blood cells, as seen through a transmission electron microscope. The black structures are nuclei.

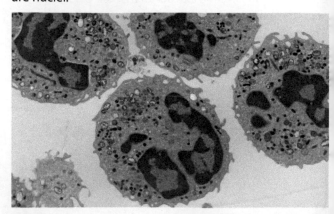

HT Here are some common units used in microscopy.

Measure	Scale	Symbol
1 metre		m
1 centimetre	$\frac{1}{100}$ th of a metre ($\times 10^{-2}$)	cm
1 millimetre	$\frac{1}{1\,000}$ (a thousandth) of a metre ($\times 10^{-3}$)	mm
1 micrometre	$\frac{1}{1\,000\,000}$ (a millionth) of a metre ($\times 10^{-6}$)	μm
1 nanometre	$\frac{1}{1\,000\,000\,000}$ (a thousand millionth or a billionth) of a metre ($\times 10^{-9}$)	nm

Comparing light and electron microscopes

Light microscope	Electron microscope
Uses light waves to produce images	Uses electrons to produce images
Low resolution	High resolution
Magnification up to ×1500	Magnification up to ×500 000 (2D) and ×100 000 (3D)
Able to observe cells and larger organelles	Able to observe small organelles
2D images only	2D and 3D images produced

Magnification and resolution

Magnification measures how many times an object under a microscope has been made larger.

To calculate the magnifying power of a microscope, use this formula:

$$\text{magnification} = \frac{\text{Size of image}}{\text{Size of real object}}$$

You may be asked to carry out calculations using magnification.

> **Example**
> A light microscope produces an image of a cell which has a diameter of 1500 µm. The cell's actual diameter is 50 µm. Calculate the magnifying power of the microscope.
>
> $\text{magnification} = \frac{1500}{50} = \times 30$

Resolution is the smallest distance between two points on a specimen that can still be told apart. The diagram below shows the limits of resolution for a light microscope. When looking through it you can see fine detail down to 200 nanometres. Electron microscopes can see detail down to 0.05 nanometres!

200 nm

Staining techniques are used in light and electron microscopy to make organelles more visible.

- **Methylene blue** is used to stain the nuclei of animal cells for viewing under the light microscope.
- **Heavy metals** such as cadmium can be used to stain specimens for viewing under the electron microscope.

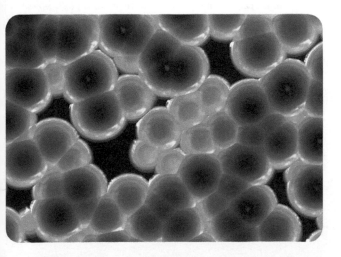

SUMMARY

- There are two main types of microscope: the light microscope and the electron microscope.
- Magnification measures how many times an object under a microscope has been enlarged.
- Resolution is the smallest distance between two points on a specimen that can still be seen apart.

QUESTIONS

QUICK TEST

1. Name the two types of microscope.

2. What is methylene blue used for when using a microscope?

3. Light microscopes can produce 2D and 3D images. True or false?

EXAM PRACTICE

1. Some bacterial colonies are grown on an agar plate. The radius of eight colonies are measured. The results are shown in the table below.

Colony	1	2	3	4	5	6	7	8
Radius/mm	10	8	2	3	2	7	5	4

Using the formula πr^2, calculate the mean cross sectional area of the colonies.

Show your working. **[3 marks]**

2. a) A student wishes to see a specimen at a higher power magnification. Which part of the light microscope must she adjust to do this? **[1 mark]**

   b) Explain why the student will not be able to observe the finer detail of organelles such as mitochondria. **[2 marks]**

   c) The same student wishes to obtain a 3D micrograph of a skin specimen surface. Which type of microscope should she use and why? **[2 marks]**

Cell division

Chromosomes

Chromosomes are found in the nucleus of eukaryotic cells. They are made of DNA and carry a large number of genes. Chromosomes exist as pairs called **homologues**.

For cells to duplicate exactly, it is important that all of the genetic material is duplicated. Chromosomes take part in a sequence of events that ensures the genetic code is transmitted precisely and appears in the new **daughter cells**.

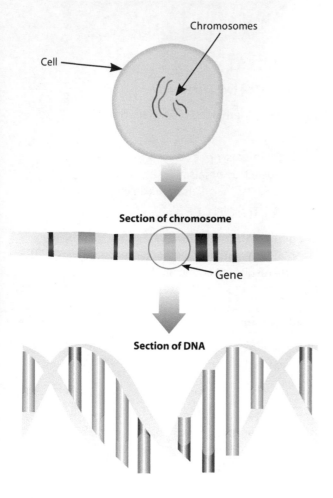

Cell

Chromosomes

Section of chromosome

Gene

Section of DNA

DNA replication

During the cell cycle, the genetic material (made of the **polymer** molecule, DNA) is doubled and then divided between the identical daughter cells. This process is called **DNA replication**.

Mitosis

Mitosis is where a **diploid** cell (one that has a complete set of chromosomes) divides to produce two more diploid cells that are genetically identical. Most cells in the body are diploid.

Humans have a diploid number of 46.

Mitosis produces new cells:
- for growth
- to repair damaged tissue
- to replace old cells
- for asexual reproduction.

Before the cell divides, the DNA is duplicated and other organelles replicate, e.g. mitochondria and ribosomes. This ensures that there is an exact copy of all the cell's content.

Mitosis – the cell copies itself to produce two genetically identical cells

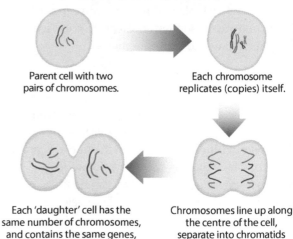

Parent cell with two pairs of chromosomes.

Each chromosome replicates (copies) itself.

Chromosomes line up along the centre of the cell, separate into chromatids and move to opposite poles.

Each 'daughter' cell has the same number of chromosomes, and contains the same genes, as the parent cell.

Meiosis

Meiosis takes place in the testes and ovaries of sexually reproducing organisms and produces gametes (eggs or sperm). The gametes are called **haploid** cells because they contain half the number of chromosomes as a diploid cell. This chromosome number is restored during **fertilisation**.

Humans have a haploid number of 23.

Meiosis – the cell divides twice to produce four cells with genetically different sets of chromosomes

Cell with two pairs of chromosomes (diploid cell).

Each chromosome replicates itself.

Cell divides for the first time.

Chromosomes part company and move to opposite poles.

Copies now separate and the second cell division takes place.

Four haploid cells (gametes), each with half the number of chromosomes of the parent cell.

Cancer

Cancer is a non-communicable disease caused by mutations in living cells.

Cancerous cells:
- divide in an uncontrolled way
- form tumours.

Benign tumours do not spread from the original site of cancer in the body. **Malignant tumour cells** invade neighbouring tissues. They spread to other parts of the body and form **secondary tumours**.

Making healthy lifestyle choices is one way to reduce the likelihood of cancer. These include:
- not smoking tobacco products (cigarettes, cigars, etc.)
- not drinking too much alcohol (causes cancer of the liver, gut and mouth)
- avoiding exposure to UV rays (e.g. sunbathing, tanning salons)
- eating a healthy diet (high fibre reduces the risk of bowel cancer) and doing moderate exercise to reduce the risk of obesity.

SUMMARY

- Multicellular organisms grow and reproduce using cell division and cell enlargement.
- Plants can also grow via differentiation into leaves, branches, etc.
- Cell division begins from the moment of fertilisation when the zygote replicates itself exactly through mitosis. Later in life, an organism may use cell division to produce sex cells (gametes) in a different type of division called meiosis.
- The different stages of cell division make up the cell cycle of an organism.

QUESTIONS

QUICK TEST

1. What name is given to sex cells?

2. Name the 46 structures in the human nucleus that carry genetic information.

3. Which structures are replicated during cell division?

EXAM PRACTICE

1. Amelia is looking at some cells through a microscope. The diagram below is a drawing of what she can see.

a) Which type of cell division is shown here? **[1 mark]**

b) Explain your answer to part a). **[2 marks]**

Metabolism – respiration

Metabolism

Metabolism is the sum of all the chemical reactions that take place in the body.

The two types of metabolic reaction are:
- building reactions (**anabolic**)
- breaking-down reactions (**catabolic**).

Anabolic Reactions

Anabolic reactions require the input of energy. Examples include:
- converting glucose to starch in plants, or glucose to glycogen in animals
- the synthesis of lipid molecules

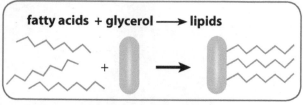

fatty acids + glycerol ⟶ lipids

- the formation of **amino acids** in plants (from glucose and nitrate ions) which, in turn, are built up into proteins.

Catabolic reactions

Catabolic reactions release energy. Examples include:
- breaking down amino acids to form **urea**, which is then excreted
- respiration.

Catabolic reactions produce waste energy in the form of heat (an **exothermic** reaction), which is transferred to the environment.

Respiration

Respiration continuously takes place in all organisms – the need to release energy is an essential life process. The reaction gives out energy and is therefore **exothermic**.

Aerobic respiration

Aerobic respiration takes place in cells. Oxygen and glucose molecules react and release energy. This energy is stored in a molecule called **ATP**.

glucose + oxygen ⟶ carbon dioxide + water + energy (locked in ATP)

HT The symbol equation for aerobic respiration is:

$$C_6H_{12}O_6 + 6O_2 \longrightarrow 6CO_2 + 6H_2O + \text{energy released}$$

Energy is used in the body for many processes, including:
- muscle contraction (for movement)
- active transport
- transmitting nerve impulses
- synthesising new molecules
- maintaining a constant body temperature.

Anaerobic respiration

Anaerobic respiration takes place in the absence of oxygen and is common in muscle cells. It quickly releases a **smaller** amount of energy than aerobic respiration through the **incomplete breakdown** of glucose.

glucose ⟶ lactic acid + energy released

In plant and yeast cells, anaerobic respiration produces different products.

glucose ⟶ ethanol + carbon dioxide + energy released

HT The symbol equation for anaerobic respiration in plant and yeast cells is:

$$C_6H_{12}O_6 \longrightarrow 2C_2H_5OH + 2CO_2 + \text{energy released}$$

This reaction is used extensively in the brewing and wine-making industries. It is also the initial process in the manufacture of spirits in a distillery.

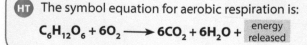

Response to exercise

In animals, anaerobic respiration takes place when muscles are working so hard that the lungs and circulatory system cannot deliver enough oxygen to break down all the available glucose through aerobic respiration. In these circumstances the **energy demand** of the muscles is high.

Anaerobic respiration and recovery

Anaerobic respiration releases energy much faster over short periods of time. It is useful when short, intense bursts of energy are required, e.g. a 100 m sprint.

However, the incomplete oxidation of glucose causes **lactic acid** to build up. Lactic acid is toxic and can cause pain, cramp and a sensation of fatigue.

The lactic acid must be broken down quickly and removed to avoid cell damage and prolonged muscle fatigue.

- During exercise the body's heart rate, breathing rate and breath volume increase so that sufficient oxygen and glucose is supplied to the muscles, and so that lactic acid can be removed.
- This continues after exercise when deep breathing or panting occurs until all the lactic acid is removed. This repayment of oxygen is called **oxygen debt**.

> **HT**
> - Lactic acid is transported to the liver where it is converted back to glucose.
> - Oxygen debt is the amount of **extra** oxygen that the body needs after exercise to react with the lactic acid and remove it from the cells.

> **WS** You may be asked to present data as graphs, tables, bar charts or histograms. For example, you could show data of breathing and heart rates on a line graph.
>
> What does the line graph below tell you about breathing and pulse rates during recovery? Why is a line graph a good way to present the data?

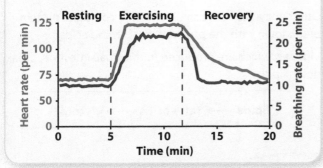

SUMMARY

- **Respiration takes place continuously in all organisms.**
- **Aerobic respiration is when oxygen and glucose molecules react and release energy.**
- **Anaerobic respiration takes place without oxygen, in muscle cells, and can cause lactic acid build up.**
- **Other organisms carrying out anaerobic respiration provide ethanol.**

QUESTIONS

QUICK TEST

1. Which type of respiration releases most energy – aerobic or anaerobic?

2. What is an exothermic reaction?

3. Give the product of anaerobic respiration in humans.

4. What type of reaction is respiration – anabolic or catabolic?

EXAM PRACTICE

1. The human muscle cell and a yeast cell can each carry out anaerobic respiration.

   Both produce a toxic waste product.

   a) Write down the word equations for both yeast anaerobic respiration and human anaerobic respiration. **[2 marks]**

   b) Compare and contrast the two types of respiration in terms of how they deal with their toxic waste products. **[3 marks]**

Metabolism – enzymes

Enzyme facts

Enzymes:

- are specific, i.e. one enzyme catalyses one reaction
- have an active site, which is formed by the precise folding of the enzyme molecule
- can be denatured by high temperatures and extreme changes in pH
- have an optimum temperature at which they work – for many enzymes this is approximately 37°C (body temperature)
- have an optimum pH at which they work – this varies with the site of enzyme activity, e.g. pepsin works in the stomach and has an optimum pH of 1.5 (acidic), salivary amylase works best at pH 7.3 (alkaline).

Enzyme activity

Enzyme molecules work by colliding with **substrate** molecules and forcing them to break up or to join with others in synthesis reactions. The theory of how this works is called the **lock and key theory**.

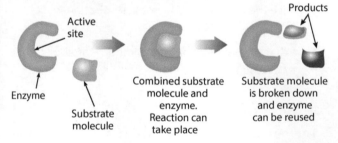

Enzyme | Active site | Substrate molecule | Combined substrate molecule and enzyme. Reaction can take place | Substrate molecule is broken down and enzyme can be reused | Products

High temperatures denature enzymes because excessive heat vibrates the atoms in the protein molecule, putting a strain on bonds and breaking them. This changes the shape of the active site.

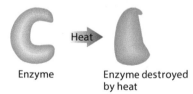

Enzyme | Heat | Enzyme destroyed by heat

In a similar way, an extreme pH alters the active site's shape and prevents it from functioning.

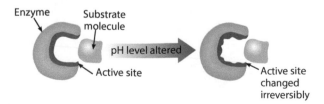

Enzyme | Substrate molecule | pH level altered | Active site | Active site changed irreversibly

At lower than the optimal temperature, an enzyme still works but much more slowly. This is because the low **kinetic energy** of the substrate and enzyme molecules lowers the number of collisions that take place. When they do collide, the energy is not always sufficient to create a bond between them.

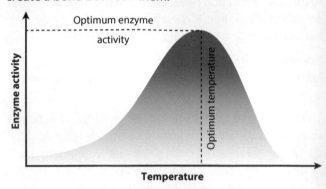

Enzymes in the digestive system

Enzymes in the digestive system help break down large nutrient molecules into smaller ones so they can be absorbed into the blood across the wall of the small intestine.

Enzyme types in the digestive system include carbohydrases, proteases and lipases.

Carbohydrases break down carbohydrates, e.g. **amylase**, which is produced in the mouth and small intestine.

> **starch ⟶ maltose**

Other carbohydrases break down complex sugars into smaller sugars.

Proteases break down protein, e.g. the enzyme pepsin, which is produced in the stomach.

> **protein ⟶ peptides ⟶ amino acids**

Other enzymes in the small intestine complete protein breakdown with the production of amino acids.

Lipases, which are produced in the small intestine, break down lipids.

> **lipids ⟶ fatty acids ⟶ glycerol**

Bile

Bile is a digestive chemical. It is produced in the liver and stored in the gall bladder.

- Its alkaline pH neutralises hydrochloric acid that has been produced in the stomach.
- It emulsifies fat, breaking it into small droplets with a large surface area.
- Its action enables lipase to break down fat more efficiently.

What happens to digested food?

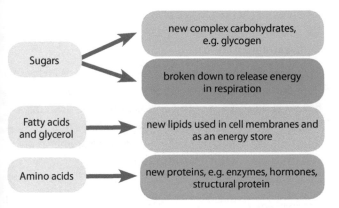

WS During your course you will investigate how certain factors affect the rate of enzyme activity. These include temperature, pH and substrate concentration.

Design an investigation to discover how temperature affects the activity of amylase. Here are some guidelines.

- Iodine solution turns from a red-brown colour to blue-black in the presence of starch.
- You can measure amylase activity by timing how long it takes for iodine solution to stop turning blue-black.
- A water bath can be set up with a thermometer. Add cold or hot water to regulate the temperature.
- Identify the independent, dependent and control variables in the investigation. Write out your method in clear steps.

SUMMARY

- Enzymes are large proteins that act as biological catalysts.
- They speed up chemical reactions, including reactions that take place in living cells, e.g. respiration, photosynthesis and protein synthesis.
- Enzymes function according to the 'lock and key' theory.

QUESTIONS

QUICK TEST

1. Which two factors affect the rate of enzyme activity?

2. What type of enzyme breaks down lipids?

3. Which molecules act as building blocks for protein polymers?

EXAM PRACTICE

1. Enzymes can only function in optimum pH and temperature conditions.

   a) Explain how a temperature of 45°C will affect the shape and function of the enzyme amylase when digesting starch. **[3 marks]**

   b) Explain how a pH of 2 allows the stomach protease enzyme to function efficiently.

      Use 'lock and key' theory to aid your explanation. **[3 marks]**

Cell transport

Diffusion

Living cells need to obtain oxygen, glucose, water, mineral ions and other dissolved substances from their surroundings. They also need to excrete waste products, such as carbon dioxide or urea. These substances pass through the cell membrane by **diffusion**.

Diffusion:

- is the (net) movement of particles in a liquid or gas from a region of high concentration to one of low concentration (down a **concentration gradient**)
- happens due to the random motion of particles past each other
- stops once the particles have completely spread out
- is passive, i.e. requires no input of energy
- can be increased in terms of rate by making the concentration gradient steeper, the diffusion path shorter, increasing the temperature or increasing the surface area over which the process occurs, e.g. having a folded cell membrane.

A protist called amoeba can absorb oxygen through diffusion.

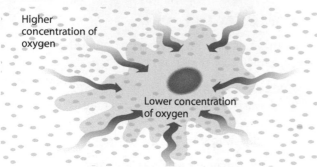

Higher concentration of oxygen

Lower concentration of oxygen

You may be asked to calculate rates of diffusion using **Fick's law**.

$$\text{rate of diffusion} \propto \frac{\text{surface area} \times \text{concentration difference}}{\text{thickness of membrane}}$$

$\propto$ means proportional to

Surface area to volume ratios

A unicellular organism (such as a protist) can absorb materials by diffusion directly from the environment. This is because it has a **large surface area to volume ratio**.

However, for a large, **multicellular organism**, the diffusion path between the environment and the inner cells of the body is long. Its large size also means that the **surface area to volume ratio** is **small**.

Adaptations

Multicellular organisms therefore need transport systems and specialised structures for exchanging materials, e.g. mammalian lungs and a small intestine, fish gills, and roots and leaves in plants. These increase diffusion efficiency in animals because they have:

- a large surface area
- a thin membrane to reduce the diffusion path
- an extensive blood supply for transport (animals)
- a ventilation system for gaseous exchange, e.g. breathing in animals.

In mammals, the individual air sacs in the lungs increase their surface area by a factor of thousands. Ventilation moves air in and out of the alveoli and the heart moves blood through the capillaries. This maintains the diffusion gradient. The capillary and alveolar linings are very thin, decreasing the diffusion path.

Single alveolus showing how increased surface area enables rapid diffusion of gases

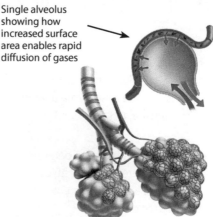

Bunches of alveoli at the end of bronchioles increase surface area (capillaries are only shown for a few alveoli)

Osmosis

Osmosis is a special case of diffusion that involves the movement of water only.

There are two ways of describing osmosis.

1. The net movement of **water** from a region of **low** solute concentration to one of **high** concentration.
2. The movement of water down a **water potential gradient**.

Osmosis:

- occurs across a **partially permeable membrane**, so solute molecules cannot pass through (only water molecules can)
- occurs in all organisms
- is passive (in other words, requires no input of energy)
- allows water movement into root hair cells from the soil and between cells inside the plant

● can be demonstrated and measured in plant cells using a variety of tissues, e.g. potato chips, daffodil stems.

Osmosis

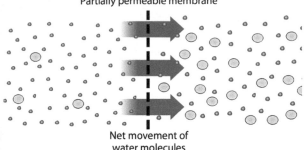

Net movement of water molecules

Active transport

Substances are sometimes absorbed against a concentration gradient, i.e. from a low to a high concentration.

Active transport:
● requires the release of energy from respiration
● takes place in the small intestine in humans, where sugar is absorbed into the bloodstream
● allows plants to absorb mineral ions from the soil through root hair cells.

A cell absorbing ions by active transport

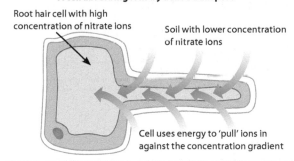

Root hair cell with high concentration of nitrate ions

Soil with lower concentration of nitrate ions

Cell uses energy to 'pull' ions in against the concentration gradient

WS During your course, you may investigate the effect of salt or sugar solutions on plant tissue.

Here is one experiment you could do.
1. Immerse raw potato cut into chips of equal length in sugar solutions of various concentration.
2. You will see the potato chips change in length depending on whether individual cells have lost or gained water.

Can you **predict** what would happen to the potato chips immersed in:
● concentrated sugar (e.g. 1 molar)
● medium concentration sugar (e.g. 0.5 molar)
● water (0 molar)?

SUMMARY

● Living cells obtain oxygen, glucose, water and other substances from their surroundings, and excrete waste products, through diffusion.
● Osmosis is a special type of diffusion involving only water.
● Active transport requires the release of energy from respiration.

QUESTIONS

QUICK TEST

1. What is Fick's Law used for?

2. Some plant tissue is placed in a highly concentrated salt solution. Explain why water leaves the cells.

EXAM PRACTICE

1. Root hair cells absorb ions from the soil and water that surround them.

   Explain how the ions can enter, despite the fact that this is against a concentration gradient. **[2 marks]**

2. A student places a thin piece of rhubarb epidermis in a strong sugar solution and observes the cells under a microscope.

   She notices that the cytoplasm has pulled away from the cell wall.

   Explain this change. **[4 marks]**

Plant tissues, organs and systems

Leaves

As this cross-section of a leaf shows, leaf tissues are adapted for efficient photosynthesis. The epidermis covers the upper and lower surfaces of the leaf and protects the plant against pathogens.

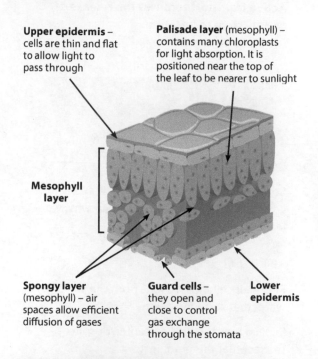

Upper epidermis – cells are thin and flat to allow light to pass through

Palisade layer (mesophyll) – contains many chloroplasts for light absorption. It is positioned near the top of the leaf to be nearer to sunlight

Mesophyll layer

Spongy layer (mesophyll) – air spaces allow efficient diffusion of gases

Guard cells – they open and close to control gas exchange through the stomata

Lower epidermis

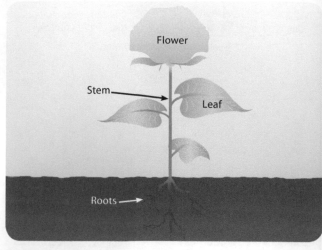

Stem and roots

Veins in the stem, roots and leaves contain tissues that transport water, carbohydrate and minerals around the plant.

● **Xylem tissue** transports water and mineral ions from the roots to the rest of the plant.

● **Phloem tissue** transports dissolved sugars from the leaves to the rest of the plant.

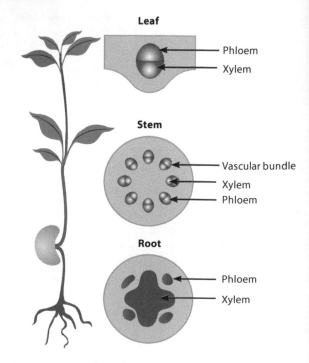

Leaf
- Phloem
- Xylem

Stem
- Vascular bundle
- Xylem
- Phloem

Root
- Phloem
- Xylem

Meristem tissue is found at the growing tips of shoots and roots.

Xylem, phloem and root hair cells

Xylem, phloem and root hair cells are adapted to their function.

Part of plant	Appearance	Function	How they are adapted to their function
Xylem	Hollow tubes made from dead plant cells (the hollow centre is called a lumen)	Transport water and mineral ions from the roots to the rest of the plant in a process called transpiration	The cellulose cell walls are thickened and strengthened with a waterproof substance called lignin
Phloem	Columns of living cells	Translocate (move) cell sap containing sugars (particularly sucrose) from the leaves to the rest of the plant, where it is either used or stored	Phloem have pores in the end walls so that the cell sap can move from one phloem cell to the next
Root hair cells	Long and thin; have hair-like extensions	Absorb minerals and water from the soil	Large surface area

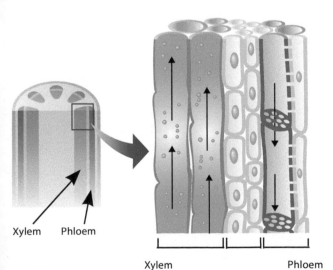

Xylem Phloem

Xylem Phloem

SUMMARY

- A plant's system is made up of organs and tissues that enable it to be a photosynthetic organism.
- Roots absorb water and minerals. They anchor plants in the soil.
- The stem transports water and nutrients to leaves. It holds leaves up to the light for maximum absorption of energy.
- The leaf is the organ of photosynthesis.
- The flower makes sexual reproduction possible through pollination.

QUESTIONS

QUICK TEST

1. What part of the plant is the main organ for photosynthesis?

2. What is the name of the waterproof substance that strengthens xylem?

EXAM PRACTICE

1. a) State one structural difference between xylem and phloem tissue. **[1 mark]**

   b) Small, herbivorous insects called aphids are found on plant stems. They have piercing mouthparts that can penetrate down to the phloem.

   Explain the reasons for this behaviour. **[2 marks]**

Transport in plants

Transpiration

The movement of water through a plant, from roots to leaves, takes place as a transpiration stream. Once water is in the leaves, it diffuses out of the stomata into the surrounding air. This is called (evapo)transpiration.

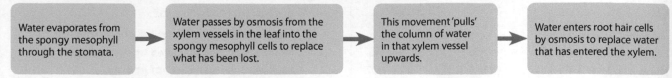

| Water evaporates from the spongy mesophyll through the stomata. | → | Water passes by osmosis from the xylem vessels in the leaf into the spongy mesophyll cells to replace what has been lost. | → | This movement 'pulls' the column of water in that xylem vessel upwards. | → | Water enters root hair cells by osmosis to replace water that has entered the xylem. |

Measuring rate of transpiration

A leafy shoot's rate of transpiration can be measured using a **potometer**.

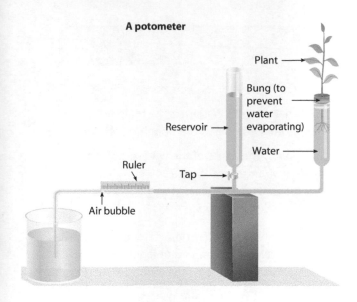

A potometer

Plant

Bung (to prevent water evaporating)

Reservoir

Water

Ruler

Tap

Air bubble

The shoot is held in a tube with a bung around the top to prevent any water from evaporating (this would give a false measurement of the water lost by transpiration) and to prevent air bubbles entering the apparatus.

As the plant transpires, it takes up water from the tube to replace what it has lost. All the water is then pulled up, moving the air bubble along.

The distance the air bubble moves can be used to calculate the plant's rate of transpiration for a given time period.

The experiment can be repeated, varying a different factor each time, to see how each factor affects the rate of transpiration.

Factors affecting rate of transpiration

Evaporation of water from the leaf is affected by **temperature**, **humidity**, **air movement** and **light intensity**.

- **Increased temperature** increases the kinetic energy of molecules and removes water vapour more quickly away from the leaf.
- **Increased air movement** removes water vapour molecules.
- **Increased light intensity** increases the rate of photosynthesis. This in turn draws up more water from the transpiration stream, which maintains high concentration in the spongy mesophyll.
- **Decreasing atmospheric humidity** lowers water vapour concentration outside of the stoma and so maintains the concentration gradient.

How water vapour exits the leaf

Sunshine

This gradient determines how quickly water vapour diffuses

Water vapour diffuses outwards

Guard cells

Opening and closing of stomata

Guard cells control the amount of water vapour that evaporates from the leaves and the amount of carbon dioxide that enters them.

- When light intensity is high and photosynthesis is taking place at a rapid rate, the sugar concentration rises in photosynthesising cells, e.g. palisade and guard cells.
- Guard cells respond to this by increasing the rate of water movement in the transpiration stream. This in turn provides more water for photosynthesis.

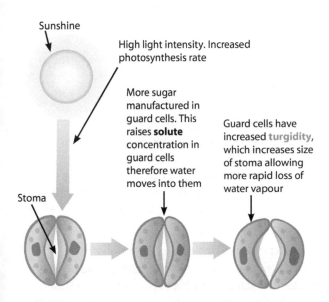

Sunshine

High light intensity. Increased photosynthesis rate

More sugar manufactured in guard cells. This raises **solute** concentration in guard cells therefore water moves into them

Guard cells have increased **turgidity**, which increases size of stoma allowing more rapid loss of water vapour

Stoma

WS A **hypothesis** is an idea or explanation that you test through study and experiments. It should include a reason. For example: desert plants have fewer stomata than temperate plants **because** they need to minimise water loss.

- In an experiment investigating the factors that affect the rate of transpiration, a student plans to take measurements of weight loss or gain from a privet plant.
- Construct hypotheses for each of these factors: **temperature**, **humidity**, **air movement** and **light intensity**.
- The first has been done for you: As temperature increases, the plant will lose mass/water more quickly **because** diffusion occurs more rapidly.

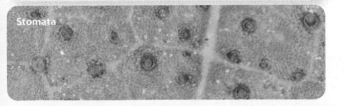

Stomata

SUMMARY

- A plant's system is made up of organs and tissues that enable it to be a photosynthetic organism.
- Roots absorb water and minerals. They anchor plants in the soil.
- The stem transports water and nutrients to leaves. It holds leaves up to the light for maximum absorption of energy.
- The leaf is the organ of photosynthesis.

QUESTIONS

QUICK TEST

1. What effect would decreasing air humidity have on transpiration?

2. Which cells control the opening and closing of the stomata?

3. What equipment is used to measure a leafy shoot's rate of transpiration?

EXAM PRACTICE

1. A student set up a plant in a potometer and noted the position of the air bubble at 2.1 cm on the scale. He left the apparatus in still air at 20°C for exactly 60 minutes. He noted the new position of the air bubble on the scale as 2.3 cm. He then immediately placed a fan close to the plant and switched it on. He noted the temperature was still 20°C. After another 60 minutes he saw that the bubble was at 5.7 cm.

   a) Calculate the rate of water absorption in cm per minute for both periods of time.

      Show your working. **[4 marks]**

   b) Explain the difference between both sets of results. **[3 marks]**

Transport in humans 1

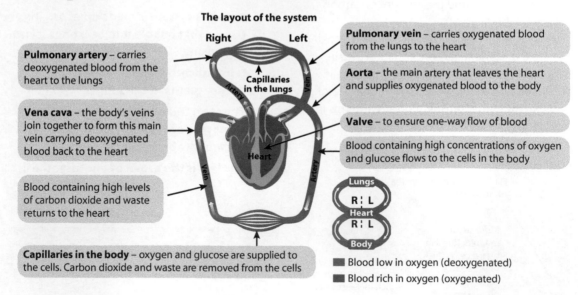

The layout of the system

Right   Left

Pulmonary artery – carries deoxygenated blood from the heart to the lungs

Pulmonary vein – carries oxygenated blood from the lungs to the heart

Capillaries in the lungs

Artery   Vein

Aorta – the main artery that leaves the heart and supplies oxygenated blood to the body

Vena cava – the body's veins join together to form this main vein carrying deoxygenated blood back to the heart

Valve – to ensure one-way flow of blood

Blood containing high concentrations of oxygen and glucose flows to the cells in the body

Heart

Vein   Artery

Blood containing high levels of carbon dioxide and waste returns to the heart

Lungs
R : L
Heart
R : L
Body

Capillaries in the body – oxygen and glucose are supplied to the cells. Carbon dioxide and waste are removed from the cells

■ Blood low in oxygen (deoxygenated)
■ Blood rich in oxygen (oxygenated)

Blood circulation

Blood moves around the body in a **double circulatory system**. In other words, blood moves twice through the heart for every full circuit. This ensures maximum efficiency for absorbing oxygen and delivering materials to all living cells.

The heart

The heart is made of powerful muscles that contract and relax rhythmically in order to continuously pump blood around the body. The **heart muscle** is supplied with nutrients (particularly glucose) and oxygen through the coronary artery.

The sequence of events that takes place when the heart beats is called the **cardiac cycle**.
- The heart relaxes and blood enters both atria from the veins.
- The atria contract together to push blood into the ventricles, opening the atrioventricular valves.
- The ventricles contract from the bottom, pushing blood upwards into the arteries. The backflow of blood into the ventricles is prevented by the **semilunar valves**.

The left side of the heart is more muscular than the right because it has to pump blood further round the body. The right side only has to pump blood to the lungs and back.

A useful measurement for scientists and doctors to take is **cardiac output**. This is calculated using:

cardiac output = stroke volume × heart rate

So, for a person who pumps out 70 ml of blood in one heartbeat (stroke volume) and has a pulse of 70 beats per minute, the cardiac output would be 4900 ml per minute.

Controlling the heartbeat

The heart is stimulated to beat rhythmically by pacemaker cells. The pacemaker cells produce impulses that spread across the atria to make them contract. Impulses are spread from here down to the ventricles, making them contract, pushing blood up and out.

Nerves connecting the heart to the brain can increase or decrease the pace of the pacemaker cells in order to regulate the heartbeat.

If a person has an irregular heartbeat, they can be fitted with an artificial, electrical pacemaker.

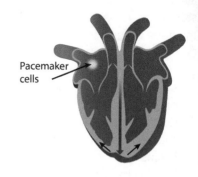

Pacemaker cells

Blood vessels

Blood is carried through the body in three types of vessel.
- **Arteries** have thick walls made of elastic fibres and muscle fibres to cope with the high pressure. The **lumen** (space inside) is small compared to the thickness of the walls. There are no valves.

- **Veins** have thinner walls. The lumen is much bigger compared to the thickness of the walls and there are valves to prevent the backflow of blood.
- **Capillaries** are narrow vessels with walls only one cell thick. These microscopic vessels connect arteries to veins, forming dense networks or **beds**. They are the only blood vessels that have permeable walls to allow the exchange of materials.

Coronary heart disease

Coronary heart disease (CHD) is a non-communicable disease. It results from the build-up of **cholesterol**, leading to plaques laid down in the coronary arteries. This restricts blood flow and the artery may become blocked with a blood clot or **thrombosis**. The heart muscle is deprived of glucose and oxygen, which causes a **heart attack**.

The likelihood of plaque developing increases if you have a high fat diet. The risk of having a heart attack can be reduced by:
- eating a balanced diet and not being overweight
- not smoking tobacco
- lowering alcohol intake
- reducing salt levels in your diet
- reducing stress levels.

Artery

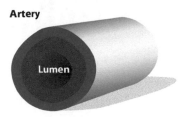

Vein

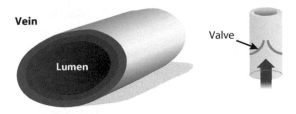

Valve

Capillary

Note: capillaries are much smaller than veins or arteries

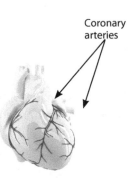

Healthy artery

Build-up of fatty material begins

Plaque forms

Plaque ruptures; blood clot forms

Coronary arteries

SUMMARY

- **Blood moves around the body in a double circulatory system.**
- **The heart is made of strong muscles that contract and relax to pump blood around the body.**
- **There are three types of blood vessels: arteries, veins and capillaries.**
- **Coronary heart disease can occur due to build-up of cholesterol in arteries.**

QUESTIONS

QUICK TEST

1. Which blood vessels contain valves?
2. Capillaries are smaller than veins. True or false?
3. Which vessel carries oxygenated blood from the lungs to the heart?

EXAM PRACTICE

1. **a)** Explain why arteries have thick, elastic muscle walls. **[1 mark]**

   b) Why do veins have valves? **[1 mark]**

2. Compared with reptiles and amphibians, humans have a double circulation.

   a) What is meant by this term? **[1 mark]**

   b) What is the advantage of having a double circulation? **[1 mark]**

Transport in humans 2

Remedying heart disease

For patients who have heart disease, artificial implants called **stents** can be used to increase blood flow through the coronary artery.

Statins are a type of drug that can be taken to reduce blood cholesterol levels.

In some people, the heart valves may deteriorate, preventing them from opening properly. Alternatively, the valve may develop a leak.

This means that the supply of oxygenated blood to vital organs is reduced. The problem can be corrected by surgical replacement using a **biological** or **mechanical** valve.

When complete heart failure occurs, a heart transplant can be carried out. If a donor heart is unavailable, the patient may be kept alive by an artificial heart until one can be found. Mechanical hearts are also used to give the biological heart a rest while it recovers.

Blood as a tissue

Blood transports digested food and oxygen to cells and removes the cells' waste products. It also forms part of the body's defence mechanism.

The four components of blood are:
- platelets
- plasma
- white blood cells
- red blood cells.

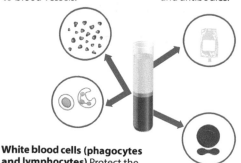

Platelets
Tiny cell fragments. They initiate the clotting process and repair damage to blood vessels.

Plasma
A straw-coloured liquid that transports dissolved food materials, urea, carbon dioxide, protein, hormones and antibodies.

White blood cells (phagocytes and lymphocytes) Protect the body against pathogens. Some have a flexible shape, which enables them to engulf invading microorganisms. Others produce antibodies.

Red blood cells (erythrocytes) Transport oxygen from lungs to tissues.

Oxygen transport

Red blood cells are small and have a biconcave shape. This gives them a large surface area to volume ratio for absorbing oxygen. When the cells reach the lungs, they absorb and bind to the oxygen in a molecule called haemoglobin.

$$\text{haemoglobin} + \text{oxygen} \rightleftharpoons \text{oxyhaemoglobin}$$

Blood is then pumped around the body to the tissues, where the reverse of the reaction takes place. Oxygen diffuses out of the red blood cells and into the tissues.

Transport of oxygen in red blood cells

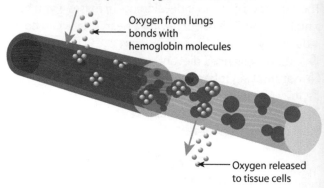

Oxygen from lungs bonds with hemoglobin molecules

Oxygen released to tissue cells

The lungs

Humans, like many vertebrates, have lungs to act as a **gaseous exchange surface**.

Other structures in the thorax enable air to enter and leave the lungs (ventilation).
- The trachea is a flexible tube, surrounded by rings of cartilage to stop it collapsing. Air is breathed in via the mouth and passes through here on its way to the lungs.
- Bronchi are branches of the trachea.
- The alveoli are small air sacs that provide a large surface area for the exchange of gases.
- Capillaries form a dense network to absorb maximum oxygen and release carbon dioxide.

In the alveoli, **oxygen** diffuses down a concentration gradient. It moves across the thin layers of cells in the alveolar and capillary walls, and into the red blood cells.

For **carbon dioxide**, the gradient operates in reverse. The carbon dioxide passes from the blood to the alveoli, and from there it travels back up the air passages to the mouth.

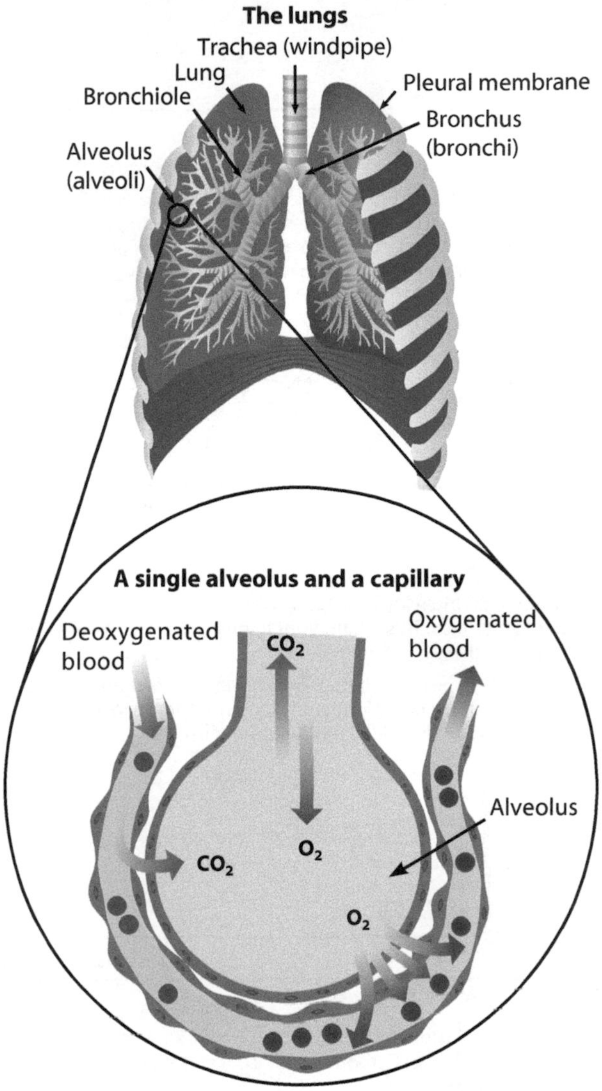

The lungs

Trachea (windpipe)
Lung
Bronchiole
Pleural membrane
Bronchus (bronchi)
Alveolus (alveoli)

A single alveolus and a capillary

Deoxygenated blood
Oxygenated blood
CO_2
CO_2
O_2
O_2
Alveolus

WS Scientists make observations, take measurements and gather data using a variety of instruments and techniques. Recording data is an important skill.

Create a table template that you could use to record data for the following experiment:

An investigation that involves measuring the resting and active pulse rates of 30 boys and 30 girls, together with their average breathing rates.

Make sure that:
- you have the correct number of columns and rows
- each variable is in a heading
- units are in the headings (so they don't need to be repeated in the body of the table).

SUMMARY

- There are four components of blood: platelets, plasma, white blood cells and red blood cells.
- Oxygen is transported around the body by red blood cells.
- The lungs act as a gaseous exchange surface.

QUESTIONS

QUICK TEST

1. Name the four components of blood.

2. What is produced when haemoglobin combines with oxygen?

EXAM PRACTICE

1. Describe the function of white blood cells. **[2 marks]**

2. A smoker's lungs develop a layer of tar lining the inside of the alveoli.

   Explain why this will make the smoker breathless.

   Use ideas about diffusion to explain your answer. **[3 marks]**

Photosynthesis

Photosynthesis:

- is an endothermic reaction
- requires chlorophyll to absorb the sunlight; this is found in the **chloroplasts** of photosynthesising cells, e.g. palisade cells, guard cells and spongy mesophyll cells
- produces **glucose**, which is then respired for energy release or converted to other useful molecules for the plant
- produces **oxygen** that has built up in the atmosphere over millions of years; oxygen is vital for respiration in all organisms.

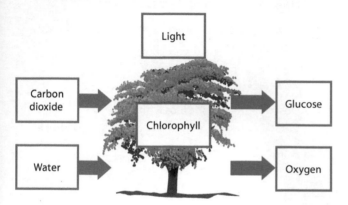

carbon dioxide + water $\xrightarrow[\text{chloroplast}]{\text{light energy}}$ glucose + oxygen

HT $6CO_2 + 6H_2O \xrightarrow[\text{chloroplast}]{\text{light energy}} C_6H_{12}O_6 + 6O_2$

Rate of photosynthesis

The rate of photosynthesis can be affected by:

- temperature
- light intensity
- carbon dioxide concentration
- amount of chlorophyll.

In a given set of circumstances, **temperature**, **light intensity** and **carbon dioxide concentration** can act as limiting factors.

Temperature

1. As the temperature rises, so does the rate of photosynthesis. This means temperature is limiting the rate of photosynthesis.
2. As the temperature approaches 45°C, the enzymes controlling photosynthesis start to be denatured. The rate of photosynthesis decreases and eventually declines to zero.

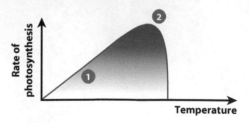

Light intensity

1. As the light intensity increases, so does the rate of photosynthesis. This means light intensity is limiting the rate of photosynthesis.
2. Eventually, the rise in light intensity has no effect on photosynthesis rate. Light intensity is no longer the limiting factor; carbon dioxide or temperature must be.

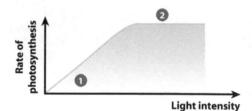

Carbon dioxide concentration

1. As carbon dioxide concentration increases, so does the rate of photosynthesis. Carbon dioxide concentration is the limiting factor.
2. Eventually, the rise in carbon dioxide concentration has no effect – it is no longer the limiting factor.

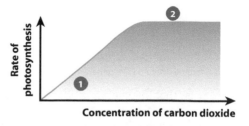

You need to understand that each factor has the potential to increase the rate of photosynthesis.

 You also need to explain how these factors interact in terms of which variable is acting as the limiting factor.

HT The inverse law

The effect of light intensity on photosynthesis can be investigated by placing a lamp at varying distances from a plant. As the lamp is moved further away from the plant the light intensity decreases, as shown in the graph.

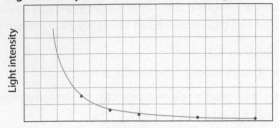

Distance from plant

There is an **inverse relationship** between the two variables. The graph can be used to convert distances to light intensity, or light intensity can be calculated using the formula:

$$\text{light intensity} = \frac{1}{d^2}$$

- d is the distance from the lamp.

This formula produces a dimensionless quantity of light intensity, so no units need to be stated. Instruments that measure absolute light intensity may use units such as 'lux' or the 'candela'.

Commercial applications

Farmers and market gardeners can increase their crop yields in greenhouses. They do this by:

- making the temperature optimum for growth using heaters
- increasing light intensity using lamps
- installing fossil-fuel burning stoves to increase carbon dioxide concentration (and increase temperature).

If applied carefully, the cost of adding these features will be offset by increased profit from the resulting crop.

Uses of glucose in plants

The glucose produced from photosynthesis can be used immediately in respiration, but some is used to synthesise larger molecules: starch, cellulose, protein and lipids.

Starch is insoluble. So it is suitable for storage in leaves, stem or roots.

Cellulose is needed for cell walls.

Protein is used for the growth and repair of plant tissue, and also to synthesise enzyme molecules.

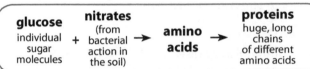

Lipids are needed in cell membranes, and for fat and oil storage in seeds.

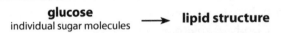

SUMMARY

- **Plants can photosynthesise, i.e. make food molecules in the form of carbohydrate from carbon dioxide and water. As such, they are the main producers of biomass.**
- **Sunlight energy is needed for photosynthesis.**
- **The rate of photosynthesis is affected by light intensity, temperature and carbon dioxide concentration.**
- **Other products such as starch and cellulose are made from glucose.**

QUESTIONS

QUICK TEST

1. Give two reasons why photosynthesis is seen as the opposite of respiration.
2. What is cellulose needed for?

EXAM PRACTICE

HT 1. The graph shows how the rate of photosynthesis changes in a plant with changing light intensity.

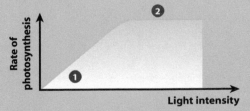

A market gardener understands this and wants to use the information to increase the yield of her tomatoes.

a) Use the graph to explain how she could accomplish this. **[4 marks]**

b) Explain why she should not invest too much money in her solution. **[2 marks]**

Non-communicable diseases

Diseases

Communicable diseases are caused by pathogens such as bacteria and viruses. They can be transmitted from organism to organism in a variety of ways. Examples include cholera and tuberculosis.

Non-communicable diseases are not primarily caused by pathogens. Examples are diseases caused by a poor diet, diabetes, heart disease and smoking-related diseases.

Health is the state of physical, social and mental well-being. Many factors can have an effect on health, including stress and life situations.

Risk factors

Non-communicable diseases often result from a combination of several **risk factors**.

- Risk factors produce an increased likelihood of developing that particular disease. They can be aspects of a person's lifestyle or substances found in the body or environment.

- Some of these factors are difficult to quantify or to establish as a definite **causal connection**. So scientists have to describe their effects in terms of probability or likelihood.

The symptoms observed in the body may result from communicable and non-communicable components interacting.

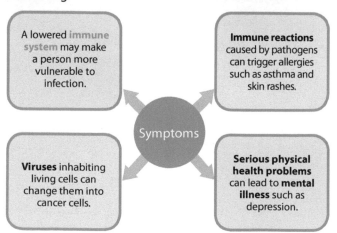

A lowered immune system may make a person more vulnerable to infection.

Immune reactions caused by pathogens can trigger allergies such as asthma and skin rashes.

Symptoms

Viruses inhabiting living cells can change them into cancer cells.

Serious physical health problems can lead to mental illness such as depression.

Poor diet

People need a **balanced diet**. If a diet does not include enough of the main food groups, malnutrition might

result. Lack of correct vitamins leads to diseases such as **scurvy** and **rickets**. Lack of the mineral, iron, results in **anaemia**. A high fat diet contributes to cardiovascular disease and high levels of salt increase blood pressure.

Smoking tobacco

Chemicals in tobacco smoke affect health.

- **Carbon monoxide** decreases the blood's oxygen-carrying capacity.

- **Nicotine** raises the heart rate and therefore blood pressure.

- **Tar** triggers cancer.

- **Particulates** cause **emphysema** and increase the likelihood of **lung infections**.

Weight/lack of exercise

Obesity and lack of exercise both increase the risk of developing **type 2 diabetes** and cardiovascular disease.

One way to show whether someone is underweight or overweight for their height is to calculate their **body mass index** (**BMI**), using the following formula:

$$BMI = \frac{mass\ (kg)}{height\ (m)^2}$$

Recommended BMI chart

BMI	What it means
<18.5	Underweight – too light for your height
18.5–25	Ideal – correct weight range for your height
25–30	Overweight – too heavy
30–40	Obese – much too heavy. Health risks!

Example

Calculate a man's BMI if he is 1.65 m tall and weighs 68 kg.

$$BMI = \frac{mass\ (kg)}{height\ (m)^2} = \frac{68}{1.65^2} = \frac{68}{2.7} = 25$$

The recommended BMI for his height (1.65 m) is 18.5–25, so he is just a healthy weight.

There are drawbacks to using BMI as a way of assessing people's health. For example:

- teenagers go through a rapid growth phase

- a person could have a well-developed muscle system – this would increase their body mass but not make them obese.

Some scientists say a more accurate method is using the waist/hip ratio. A tape measure is used to measure the circumference of the hips and the waist (at its widest). The waist measurement is divided by the hips measurement. The following chart can then be used.

Waist to hip ratio (WHR)		
Male	Female	Health risk based solely on WHR
0.95 or below	0.80 or below	Low
0.96 to 1.0	0.81 to 0.85	Moderate
1.0+	0.85+	High

Alcohol

Drinking excess alcohol can impair brain function and lead to **cirrhosis** of the liver. It also contributes to some types of cancer and cardiovascular disease.

Smoking and drinking alcohol during pregnancy

Unborn babies receive nutrition from the mother via the placenta. Substances from tobacco, alcohol and other drugs can pass to the baby and cause **lower birth weight**, **foetal alcohol syndrome** and **addiction**.

Carcinogens

Exposure to **ionising radiation** (e.g. X-rays, gamma rays) can cause cancerous tumours. Overexposure to UV light can cause skin cancer. Certain chemicals such as mercury can also increase the likelihood of cancer.

(WS) Interpreting complex data in graphs doesn't need to be difficult. This line graph shows data about smoking and lung cancer. Look for different patterns in it. For example:

- males have higher smoking rates in all years
- female cancer rates have increased overall since 1972.

Can you see any other patterns?

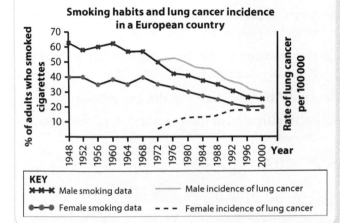

SUMMARY

- Diseases can be communicable or non-communicable.
- Communicable diseases are caused by pathogens. Non-communicable diseases include heart disease, diabetes and smoking-related illnesses.
- Symptoms of diseases are often due to an interaction between communicable and non-communicable factors.
- Risk factors include poor diet, smoking tobacco, lack of exercise and alcohol abuse.

QUESTIONS

QUICK TEST

1. What does BMI stand for?

2. What is a pathogen?

EXAM PRACTICE

1. Write down two conditions that are non-communicable. [2 marks]

2. a) A woman's height is 1.55m and she weighs 64kg. What is her BMI? [2 marks]

   b) Using the BMI table on p.30, state whether you think she should take steps to lose weight.

   Give a reason for your answer. [2 marks]

Communicable diseases

How do pathogens spread?

Pathogens are disease-causing microorganisms from groups of bacteria, viruses, fungi and protists. All animals and plants can be affected by pathogens. They spread in many ways, including:

- **droplet infection** (sneezing and coughing), e.g. flu
- **physical contact**, such as touching a contaminated object or person, e.g. Ebola virus
- **transmission** by transferral of or contact with bodily fluids, e.g. hepatitis B
- **sexual transmission**, e.g. HIV, gonorrhoea
- **contamination of food or water**, e.g. Salmonella, cholera
- **animal bites**, e.g. rabies.

How do pathogens cause harm?

- Bacteria and viruses reproduce rapidly in the body.
- Viruses cause cell damage.
- Bacteria produce toxins that damage tissues.

These effects produce symptoms in the body.

How can the spread of disease be prevented?

The spread of disease can be prevented by:

- good hygiene, e.g. washing hands/whole body, using soaps and disinfectants
- destroying vectors, e.g. disrupting the life cycle of mosquitoes can combat malaria
- the isolation or quarantine of individuals
- vaccination.

Bacterial diseases

Disease	Transmission	Symptoms	Treatment/prevention
Tuberculosis	Droplet infection	Persistent coughing, which may bring up blood; chest pain; weight loss; fatigue; fever; night sweats; chills	Long course of antibiotics
Cholera	Contaminated water/food	Diarrhoea; vomiting; dehydration	Rehydration salts
Chlamydia	Sexually transmitted	May not be present, but can include discharge and bleeding from sex organs	Antibiotics. Using condoms during sexual intercourse can reduce chances of infection
Helicobacter	Spread orally from person to person by saliva, or by fecal contamination	Abdominal pain; feeling bloated; nausea; vomiting; loss of appetite; weight loss	Proton pump inhibitors and antibiotics
Salmonella	Contaminated food containing toxins from pathogens – these could be introduced from unhygienic food preparation techniques	Vomiting; fever; diarrhoea, stomach cramps	Anti-diarrhoeals and antibiotics; vaccinations for chickens
Gonorrhoea	Sexually transmitted	Thick yellow or green discharge from vagina or penis; pain on urination	Antibiotic injection followed by antibiotic tablets; penicillin is no longer effective against gonorrhoea; prevention through use of condoms

Fungal diseases

Disease	Transmission	Symptoms	Treatment/prevention
Athlete's foot	Direct and indirect contact, e.g. skin-to-skin, bed sheets and towels (often spreads at swimming pools and in changing rooms)	Itchy, red, scaly, flaky and dry skin	Self-care and anti-fungal medication externally applied

Viral and protist diseases

Disease	Transmission	Symptoms and notes	Treatment/prevention
Measles (viral)	Droplets from sneezes and coughs	Fever; red skin rash; fatal if complications arise	No specific treatment; vaccine is a highly effective preventative measure
HIV (viral)	Sexually transmitted; exchange of body fluids; sharing of needles during drug use	Flu-like symptoms initially; late-stage AIDS produces complications due to compromised immune system	Anti-retroviral drugs
Ebola (viral)	Via body fluids; contaminated needles; bite from infected animal	Early stages: muscle pain; sore throat; diarrhoea Later stages: kidney/liver failure; internal bleeding	No direct treatments or tested vaccines are available at the time of writing; symptoms and infections are treated as they appear
Malaria (protist)	Via mosquito vector	Headache; sweats; chills and vomiting; symptoms disappear and reappear on a cyclical basis; further life-threatening complications may arise	Various anti-malarial drugs are available for both prevention and cure; prevention of mosquito breeding and use of mosquito nets

Malarial parasite

The **plasmodium** is a **protist** that causes malaria. It can reproduce asexually in the human host but sexually in the mosquito.

Mosquito

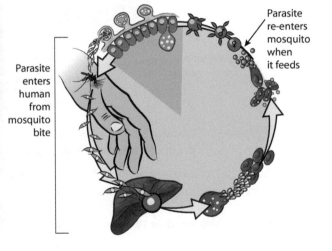

Parasite enters human from mosquito bite

Parasite re-enters mosquito when it feeds

SUMMARY

- **Pathogens can be spread through physical contact, droplet infection, sexual transmission, contamination of food/water, transmission of bodily fluids, or animal bites.**
- **Bacterial diseases include salmonella and gonorrhoea.**
- **Viral and protist diseases include measles, HIV and malaria.**

QUESTIONS

QUICK TEST

1. Why is cholera transmitted rapidly in areas that have poor sanitation?
2. Name a fungal disease.

EXAM PRACTICE

1. Microorganisms consist of bacteria, viruses, fungi and Protista.

   Many cause harm to the human body.

   a) Write down the term that describes such organisms. **[1 mark]**

   b) Harmful microorganisms produce symptoms when they reproduce in large numbers.

   Write down **two** ways in which microorganisms do this. **[2 marks]**

2. Malaria kills many thousands of people every year. The disease is common in areas that have warm temperatures and stagnant water.

   Explain why this is. **[2 marks]**

Human defences

Non-specific defences

The body has a number of general or non-specific defences to stop pathogens multiplying inside it.

The skin covers most of the body – it is a **physical barrier** to pathogens. It also secretes antimicrobial peptides to kill microorganisms. If the skin is damaged, a clotting mechanism takes place in the blood preventing pathogens from entering the site of the wound.

Tears contain enzymes called **lysozymes**. Lysozymes break down pathogen cells that might otherwise gain entry to the body through tear ducts.

Hairs in the nose trap particles that may contain pathogens.

Tubes in the respiratory system (**trachea** and **bronchi**) are lined with special epithelial cells. These cells either produce a sticky, liquid mucus that traps microorganisms or have tiny hairs called cilia that move the mucus up to the mouth where it is swallowed.

The stomach produces **hydrochloric acid**, which kills microorganisms.

Phagocytes are a type of white blood cell. They move around in the bloodstream and body tissues searching for pathogens. When they find pathogens, they engulf and digest them in a process called **phagocytosis**.

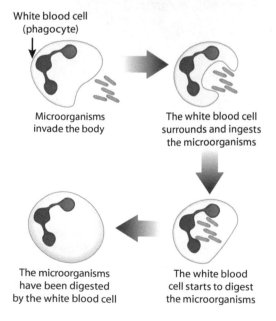

White blood cell (phagocyte)

Microorganisms invade the body

The white blood cell surrounds and ingests the microorganisms

The white blood cell starts to digest the microorganisms

The microorganisms have been digested by the white blood cell

Specific defences

White blood cells called **lymphocytes** recognise molecular markers on pathogens called antigens. They produce antibodies that lock on to the antigens on the cell surface of the pathogen cell. The immobilised cells are clumped together and engulfed by phagocytes.

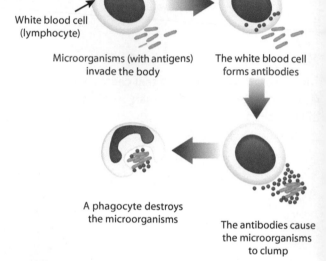

White blood cell (lymphocyte)

Microorganisms (with antigens) invade the body

The white blood cell forms antibodies

A phagocyte destroys the microorganisms

The antibodies cause the microorganisms to clump

Some white blood cells produce **antitoxins** that neutralise the poisons produced from some pathogens.

Every pathogen has its own unique antigens. Lymphocytes make antibodies specifically for a particular antigen.

Example: Antibodies to fight TB will not fight cholera

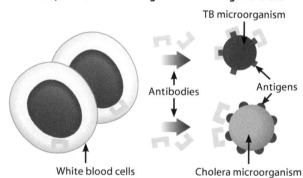

TB microorganism

Antibodies

Antigens

White blood cells

Cholera microorganism

Active immunity

Once lymphocytes recognise a particular pathogen, the interaction is stored as part of the body's immunological memory through memory lymphocytes. These memory cells can produce the right antibodies much quicker if the same pathogen is detected again, therefore providing future protection against the disease. The process is called the secondary response and is part of the body's active immunity.

Active immunity can also be achieved through vaccination.

Memory lymphocytes and antibody production

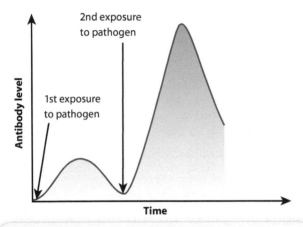

WS Investigating the growth of microorganisms involves culturing microorganisms.

This presents hazards that require a risk assessment. A risk assessment involves taking into account the severity of each hazard and the likelihood that it will occur.

Any experiment of this type involves thinking about risks in advance. Here is an example of a risk assessment table for this investigation.

Hazard	Infection from pathogen	Scald from autoclave (a specialised pressure cooker for superheating its contents)
Risk	High	High
How to lower the risk	● Observe aseptic technique. ● Wash hands thoroughly before and after experiment. ● Store plates at a maximum temperature of 25°C.	● Ensure lid is tightly secured. ● Adjust heat to prevent too high a pressure. ● Wait for autoclave to cool down before removing lid.

SUMMARY

● The body has a number of non-specific defences, including the skin, tears, and stomach acid.

● The body has specific defences in the form of lymphocytes.

● Lymphocytes produce antibodies, which lock on to antigens.

● Memory cells remember the shape of antigens and provide a defence against further infection. This is active immunity.

QUESTIONS

QUICK TEST

1. Describe the process of phagocytosis.

2. What is an antigen?

3. What enzyme is found in tears?

EXAM PRACTICE

1. The diagram shows a white blood cell producing small proteins as part of the body's immune system.

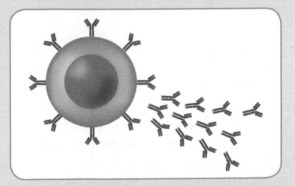

a) What is the name of these proteins? **[1 mark]**

b) i) What is the name of this type of white blood cell? **[1 mark]**

   ii) Describe what happens to these components if the body is invaded again by the same pathogen. **[3 marks]**

Fighting disease

Vaccination

There are two types of vaccination: active and passive.

Passive immunisation

Antibodies are introduced into an individual's body, rather than the person producing them on their own. Some pathogens or toxins (e.g. snake venom) act very quickly and a person's immune system cannot produce antibodies quickly enough. So the person must be injected with the antibodies. However, this does not give long-term protection.

Active immunisation

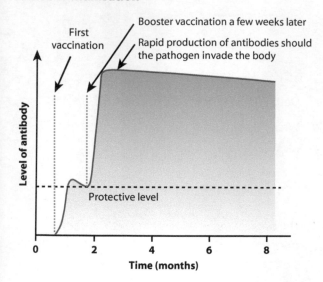

Benefits of immunisation	Risks of immunisation
● It protects against diseases that could kill or cause disability (e.g. polio, measles). ● If everybody is vaccinated and herd immunity is established, the disease eventually dies out (this is what happened to smallpox).	● A person could have an allergic reaction to the vaccine (small risk).

Immunisation gives a person immunity to a disease without the pathogens multiplying in the body, or the person having symptoms.

- A weakened or inactive strain of the pathogen is injected. The pathogen is heat-treated so it cannot multiply. The antigen molecules remain intact.

- Even though they are harmless, the antigens on the pathogen trigger the white blood cells to produce specific antibodies.

- As with natural immunity, **memory lymphocytes** remain sensitised. This means they can produce more antibodies very quickly if the same pathogen is detected again.

Antibiotics and painkillers

Diseases caused by bacteria (not viruses) can be treated using **antibiotics**, e.g. penicillin. Antibiotics are drugs that destroy the pathogen. Some bacteria need to be treated with antibiotics specific to them.

Antibiotics work because they **inhibit** cell processes in the bacteria but not the body of the host.

Viral diseases can be treated with **antiviral drugs**, e.g. swine flu can be treated with 'Tamiflu' tablets. It is a challenge to develop drugs that destroy viruses without harming body tissues.

Antibiotic resistance

Antibiotics are very effective at killing bacteria. However, some bacteria are **naturally resistant** to particular antibiotics. It is important for patients to follow instructions carefully and take the full course of antibiotics so that all the harmful bacteria are killed.

If doctors over-prescribe antibiotics, there is more chance of resistant bacteria surviving. These multiply and spread, making the antibiotic useless. **MRSA** is a bacterium that has become resistant to most antibiotics. These bacteria have been called 'superbugs'.

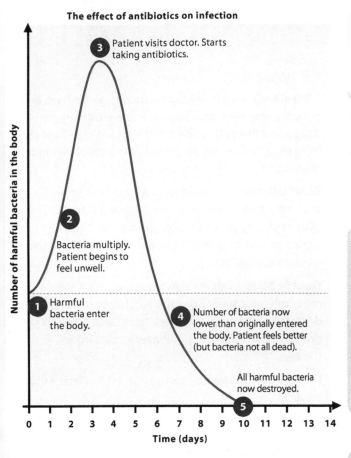

The effect of antibiotics on infection

Number of harmful bacteria in the body

3 Patient visits doctor. Starts taking antibiotics.

2 Bacteria multiply. Patient begins to feel unwell.

1 Harmful bacteria enter the body.

4 Number of bacteria now lower than originally entered the body. Patient feels better (but bacteria not all dead).

All harmful bacteria now destroyed.

5

Time (days): 0 1 2 3 4 5 6 7 8 9 10 11 12 13 14

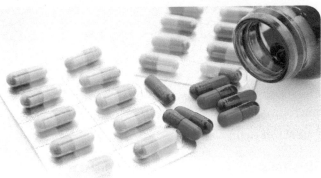

Painkillers

Painkillers or **analgesics** are given to patients to relieve symptoms of a disease, but they do not kill pathogens. Types of painkiller include paracetamol and ibuprofen. Morphine is another painkiller – it is a medicinal form of heroin used to treat extreme pain.

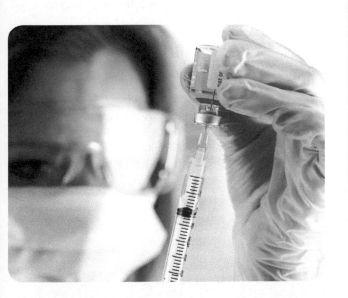

SUMMARY

- There are two types of vaccination: passive immunisation and active immunisation.
- Diseases caused by bacteria can be treated using antibiotics, but some bacteria are resistant to antibiotics.
- Painkillers do not kill pathogens; they only relieve symptoms.

QUESTIONS

QUICK TEST

1. What is an antibiotic?

2. Name the two types of vaccination.

3. What can doctors and patients do to reduce the risk of antibiotic-resistant bacteria developing?

EXAM PRACTICE

1. Explain how a vaccine works. **[4 marks]**

2. Bella is suffering from the flu and has asked her doctor for some antibiotics.

   a) How do antibiotics work? **[1 mark]**

   b) Explain why the doctor will not prescribe her antibiotics. **[2 marks]**

3. Explain why antibiotics are becoming increasingly less effective against 'superbugs' such as MRSA. **[3 marks]**

Discovery and development of drugs

Discovery of drugs

The following drugs are obtained from plants and microorganisms.

Name of drug	Use
Digitalis Foxgloves (common garden plants that are found in the wild)	Slows down the heartbeat; can be used to treat heart conditions
Aspirin From Willow trees (aspirin contains the active ingredient salicylic acid)	Mild painkiller
Penicillin Penicillium mould (discovered by Alexander Fleming)	Antibiotic

Modern pharmaceutical drugs are synthesised by chemists in laboratories, usually at great cost. The starting point might still be a chemical extracted from a plant.

New drugs have to be developed all the time to combat new and different diseases. This is a lengthy process, taking up to ten years. During this time the drugs are tested to determine:

- that they work
- that they are safe
- that they are given at the correct **dose** (early tests usually involve low doses).

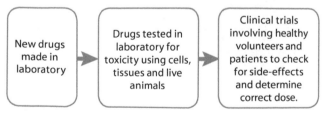

New drugs made in laboratory → Drugs tested in laboratory for toxicity using cells, tissues and live animals → Clinical trials involving healthy volunteers and patients to check for side-effects and determine correct dose.

In addition to testing, **computer models** are used to predict how the drug will affect cells, based on knowledge about how the body works and the effects of similar drugs. There are many who believe this type of testing should be extended and that animal testing should be phased out.

Clinical trials

Clinical trials are carried out on healthy volunteers and patients who have the disease. Some are given the new drug and others are given a placebo. The effects of the drug can then be compared to the effects of taking the placebo.

Blind trials involve volunteers who do not know if they have been given the new drug or a placebo. This eliminates any psychological factors and helps to provide a fair comparison. (Blind trials are not normally used in modern clinical trials.)

Double blind trials involve volunteers who are randomly allocated to groups. Neither they nor the doctors/scientists know if they have been given the new drug or a placebo. This eliminates all bias from the test.

New drugs must also be tested against the best existing treatments.

WS When studies involving new drugs are published, there is a **peer review**. This is where scientists with appropriate knowledge read the scientific study and examine the data to see if it is **valid**. Sometimes the trials are duplicated by others to see if similar results are obtained. This increases the **reliability** of the findings and filters out false or exaggerated claims.

Once a **consensus** is agreed, the paper is published. This allows others to hear about the work and to develop it further.

In the case of pharmaceutical drugs, clinical bodies have to decide if the drug can be **licensed** (allowed to be used) and whether it is **cost-effective**. This can be controversial because a potentially life-saving drug may not be used widely simply because it costs too much and/or would benefit too few people.

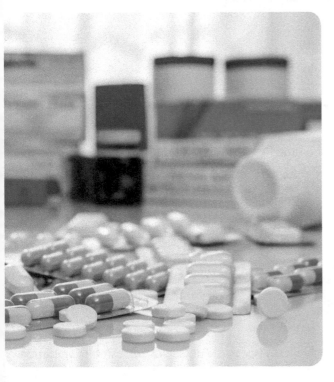

- Drugs used for treating illnesses and health conditions include antibiotics, analgesics and other chemicals that modify body processes and chemical reactions.
- In the past, these drugs were obtained from plants and microorganisms.
- Clinical trials are carried out to test new drugs.

QUESTIONS

QUICK TEST

1. What is a blind trial?

2. The drug, Digitalis, is obtained from foxglove plants. True or false?

EXAM PRACTICE

1. **a)** New pharmaceutical drugs have to be extensively tested before their release on to the market.

   Explain why this is. **[2 marks]**

   b) Name one alternative to animal testing of drugs. **[1 mark]**

HT 2. Scientists can work out how dangerous a certain drug is by calculating a therapeutic ratio. They use the following formula:

$$\text{Therapeutic ratio} = \frac{\text{Lethal dose}}{\text{Smallest dose needed to have an effect}}$$

The larger the ratio, the safer it is.

A new painkilling drug has been approved by the Medicines Agency. It has a lethal dose of 100 mg for a 100 kg man. The smallest dose needed to have an effect on the same man is 5 mg.

a) Calculate the therapeutic ratio. Show your working. **[2 marks]**

b) Morphine, another painkiller, has a therapeutic ratio of 6.

Use your answer to part a) to judge whether the new painkilling drug is more or less dangerous than morphine. **[1 mark]**

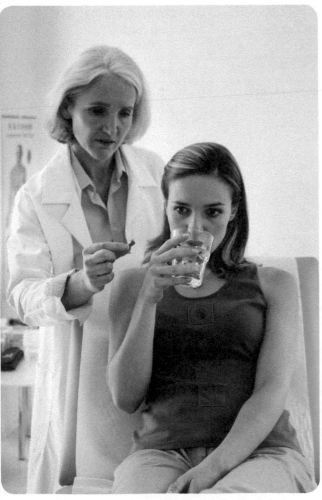

Plant diseases

Pathogens and pests that affect plants

Disease	Pathogen	Appearance/effect on plants	Treatment
Rose black spot	Fungal disease – the fungal spores are spread by water and wind	Purple/black spots on leaves; these then turn yellow and drop early, leading to a lack of photosynthesis and poor growth	Apply a fungicide and/or remove affected leaves Don't plant roses too close together. Avoid wetting leaves
Tobacco mosaic virus (TMV)	Widespread disease that affects many plants (including tomatoes)	'Mosaic' pattern of discolouration; can lead to lack of photosynthesis and poor growth	Remove infected plants Crop rotation Wash hands after treating plant
Ash dieback	Caused by the fungus *Chalara*	Leaf loss and bark lesions	Cut back or remove diseased trees to reduce chance of airborne spores
Barley powdery mildew	*Erysiphe graminis*	Causes powdery mildew to appear on grasses, including cereals	Fungicides and careful application of nitrogen fertilisers
Crown gall disease	*Agrobacterium tumefaciens*	Tumours or 'galls' at the crown of plants such as apple, raspberry and rose	Use of copper and methods of biological control

Pests	What they do	Appearance/effect on plants	Control
Invertebrates and particularly insects, e.g. many species of aphids	Feed on sap, leaves and storage organs; transmit pathogenic viruses		Chemical pesticides or biological control methods

Mineral ion deficiencies

Plants need **mineral ions** to build complex molecules. The ions are obtained from the soil via the roots in an active manner (requiring energy). In particular, plants need:

- **nitrates** to form **amino acids**, the building blocks of **proteins**. They are also needed to make nucleic acids such as **DNA**. Lack of nitrates in a plant leads to yellow leaves and stunted growth

- **magnesium** to form chlorophyll, which absorbs light energy for photosynthesis. Lack of magnesium results in chlorosis, which is a discolouration of the leaves.

Defence responses of plants

Plants defend themselves from herbivores and disease in a range of ways. These adaptations have evolved over millions of years to maximise a plant's survival.

Mechanical defences include:

- thorns and hairs to deter plant-eaters

- leaves that droop or curl on contact

- mimicry, to fool animals into avoiding them as food or laying their eggs on them. For example, passion flowers have structures that look like yellow eggs. A butterfly is less likely to lay eggs on a plant that has apparently already been used.

Physical defences include:

- tough, waxy leaf cuticles

- cellulose cell walls

- layers of dead cells around stems, e.g. bark. These fall off, taking pathogens with them.

Chemical defences include:

- producing antibacterial chemicals, e.g. mint and witch hazel

- producing toxins to deter herbivores, e.g. deadly nightshade, foxgloves and tobacco plants.

SUMMARY

- Pathogens can affect plants. Diseases include Rose black spot and Tobacco mosaic virus (TMV).
- Plants defend themselves from pests and pathogens in three ways: mechanical defences, physical defences and chemical defences.

QUESTIONS

QUICK TEST

1. What kind of disease is Rose black spot?

2. How can Ash dieback be treated?

3. How can barley powdery mildew be treated?

4. What do pests feed on when on a plant?

EXAM PRACTICE

1. Blackspot is a fungal disease that affects rose leaves.

   Fatima has noticed that roses growing in areas with high air pollution are less affected by the disease.

   She thinks that regular exposure to acid rain is stopping the disease from developing.

   Describe a simple experiment that Fatima could do to test her theory. **[3 marks]**

2. State one mechanical and one physical defence that plants use to defend themselves against herbivores or disease. **[2 marks]**

Homeostasis and negative feedback

Homeostasis

The body has automatic control systems to maintain a constant internal environment (**homeostasis**). These systems make sure that cells function efficiently.

Homeostasis balances inputs and outputs to ensure that optimal levels of temperature, pH, water, oxygen and carbon dioxide are maintained.

For example, even in the cold, homeostasis ensures that body temperature is regulated at about 37°C.

Control systems in the body may involve the nervous system, the endocrine system, or both. There are three components of control.

- **Effectors** cause responses that restore optimum levels, e.g. muscles and glands.
- **Coordination centres** receive and process information from the receptors, e.g. brain, spinal cord and pancreas.
- **Receptors** detect stimuli from the environment, e.g. taste buds, nasal receptors, the inner ear, touch receptors and receptors on retina cells.

WS When taking measurements, the quality of the measuring instrument and a scientist's skill is very important to achieve **accuracy**, **precision** and **minimal error**.

Adrenaline levels in blood plasma are measured by a chromatography method called HPLC. This is often coupled to a detector that gives a digital readout.

This digital readout displays the concentration of adrenaline as 6.32. This means that the instrument is precise up to $\frac{1}{100}$ of a unit.

A less precise instrument might only measure down to $\frac{1}{10}$ of a unit, e.g. 6.3 (one decimal place).

A bar graph of some data generated from HPLC is shown alongside. The graph shows the **average concentration** of three different samples of blood. The average is taken from many individual measurements.

The vertical error bars indicate the range of measurements (the difference between highest and lowest) obtained for each sample.

Sample C shows **the greatest precision** as the individual readings do not vary as much as the others. There is less error so we can be more confident that the average is closer to the **true value** and therefore more **accurate**.

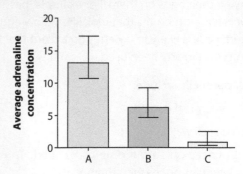

HT Negative feedback

Negative feedback occurs frequently in homeostasis. It involves the automatic reversal of a change in the body's condition.

In the body, examples of negative feedback include osmoregulation/water balance, balancing blood sugar levels, maintaining a constant body temperature and controlling metabolic rate.

Metabolism needs to be controlled so that chemical reactions in the body take place at an optimal rate. Negative feedback controls metabolic rate by using the hormones **thyroxine** and **adrenaline**.

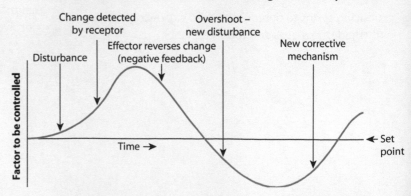

HT Thyroxine

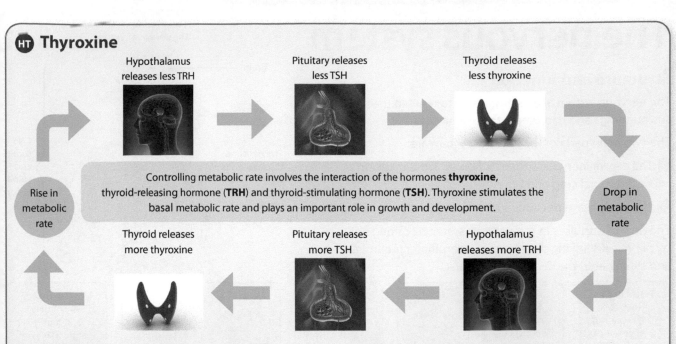

Hypothalamus releases less TRH → Pituitary releases less TSH → Thyroid releases less thyroxine → Drop in metabolic rate

Controlling metabolic rate involves the interaction of the hormones **thyroxine**, thyroid-releasing hormone (**TRH**) and thyroid-stimulating hormone (**TSH**). Thyroxine stimulates the basal metabolic rate and plays an important role in growth and development.

Rise in metabolic rate ← Thyroid releases more thyroxine ← Pituitary releases more TSH ← Hypothalamus releases more TRH

Adrenaline

Adrenaline is sometimes called the 'flight or fight' hormone. During times of stress the adrenal glands produce adrenaline. It has a direct effect on muscles, the liver, intestines and many other organs to prepare the body for sudden bursts of energy. Specifically, adrenaline increases the heart rate so that the brain and muscles receive oxygen and glucose more rapidly.

SUMMARY

- **The body has control systems that work automatically to maintain a constant internal environment. This is homeostasis.**
- **Negative feedback is the automatic reversal of a change in the body's condition.**

QUESTIONS

QUICK TEST

1. At what temperature is the human body regulated?
HT 2. Give two examples where negative feedback operates in the human body.

QUESTIONS

EXAM PRACTICE

HT 1. Match the organs below, with the function they perform:

Pancreas	Releases TSH
Skin receptor	Detects pressure
Pituitary gland	Releases TRH
Hypothalamus	Produces insulin

[3 marks]

The nervous system

Structure and function

The nervous system allows organisms to react to their surroundings and coordinate their behaviour.

The two main parts of the nervous system are:

- the central nervous system (**CNS**), which is made up of the spinal cord and brain

- the **peripheral nervous system**.

The flow of impulses in the nervous system is carried out by nerve cells linking the receptor, coordinator (neurones and synapses in the CNS) and effector.

The main components of the nervous system

Brain

Spinal cord

The neurones that make up the peripheral nervous system

CNS (brain and spinal cord)

Sense organ	Sensory neurone	Synapse	Relay neurone	Synapse	Motor neurone	Muscle
In the sense organ, receptors detect a change – either inside or outside the body. The change is a stimulus.	Conducts the impulse from the sense organ towards the CNS.	The gap between the sensory and relay neurones.	Passes the impulse on to a motor neurone.	The gap between the relay neurone and the motor neurone.	Passes the impulse on to the muscle (or gland).	The muscle responds by contracting, which results in a movement. Muscles and glands are examples of effectors.

Nerve cells or neurones are specially adapted to carry nerve impulses, which are electrical in nature. The impulse is carried in the long, thin part of the cell called the axon.

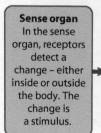
Motor neurone

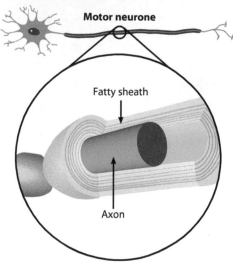

Fatty sheath

Axon

There are three types of neurone:

Sensory neurones carry impulses from receptors to the CNS.

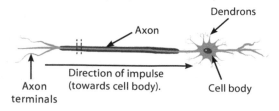

Dendrons

Axon

Axon terminals

Direction of impulse (towards cell body).

Cell body

Relay neurones make connections between neurones inside the CNS.

Dendron

Axon terminals

Impulse travels first towards, and then away from, cell body.

Motor neurones carry impulses from the CNS to muscles and glands.

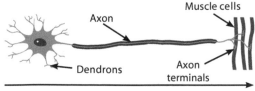

Axon

Muscle cells

Dendrons

Axon terminals

Direction of impulse (away from cell body).

Synapses

Synapses are junctions between neurones.

They play an important part in regulating the way impulses are transmitted. Synapses can be found between different neurones, neurones and muscles, and between dendrites (the root-like outgrowths from the cell body).

When an impulse reaches a synapse, a neurotransmitter is released by the neurone ('A' in the following diagram) into the gap that lies between the neurones. It travels by diffusion and binds to receptor molecules on the next neurone. This triggers a new electrical impulse to be released.

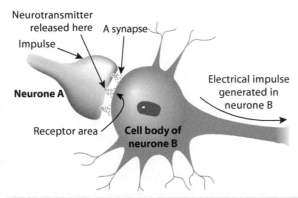

Neurotransmitter released here
A synapse
Impulse
Neurone A
Electrical impulse generated in neurone B
Receptor area
Cell body of neurone B

WS You may have to investigate the effect of factors on human reaction time.

For example, you could be asked to investigate a learned reflex by measuring how far up a ruler someone can catch it. The nearer to the zero the ruler is caught, the faster the reflex.

You could investigate factors such as:
● experience/practice at catching
● sound ● touch ● sight.

Can you design experiments to test these variables? Which factors will need to be kept the same?

SUMMARY

● **The nervous system is made up of the brain and spinal cord (CNS) and neurones.**

● **There are three types of neurone: sensory, relay and motor neurones.**

● **Synapses are junctions between neurones where a neurotransmitter is released.**

● **A reflex arc is the pathway taken by impulses around the body in response to a reflex action.**

QUESTIONS

QUICK TEST

1. Name the junctions between neurones.

2. How do reflex arcs aid survival of an organism?

Reflex arcs

Reflex actions:
● are involuntary/automatic
● are very rapid
● protect the body from harm
● bypass conscious thought.

The pathway taken by impulses around the body is called a reflex arc. Examples include:
● opening and closing the pupil in the eye
● the knee-jerk response
● withdrawing your hand from a hot plate.

Here is the arc pathway for a pain response.

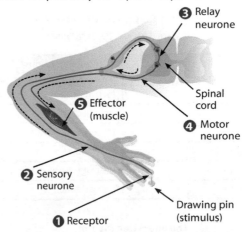

❸ Relay neurone
Spinal cord
❺ Effector (muscle)
❹ Motor neurone
❷ Sensory neurone
Drawing pin (stimulus)
❶ Receptor

QUESTIONS

EXAM PRACTICE

1. The flow chart illustrates the events that occur during a reflex action.

   X → **Receptor** → **Sensory neurone** → **Relay neurone** → **Brain and/or spinal cord** → **Motor neurone** → **Effector**

   Paul accidentally puts his hand on a pin. Without thinking, he immediately pulls his hand away.

   a) Which component of a reflex arc is represented by the letter X? **[1 mark]**

   b) Give **two** reasons why this can be described as a reflex action. **[2 marks]**

   c) Use the features in the flow chart to describe what happens in this reflex action. **[4 marks]**

The endocrine system

Structure and function

The endocrine system is made up of glands that are ductless and secrete **hormones** directly into the bloodstream. The blood carries these chemical messengers to **target organs** around the body, where they cause an effect.

Hormones:

- are large protein molecules
- interact with the nervous system to exert control over essential biological processes
- act over a longer time period than nervous responses but their effects are slower to establish.

Endocrine gland	Hormone(s) produced
Pituitary gland	TSH, ADH, FSH, LH, etc.
Pancreas	Insulin, glucagon
Thyroid	Thyroxine
Adrenal gland	Adrenaline
Ovaries (female)	Oestrogen, progesterone
Testes (male)	Testosterone

The endocrine system

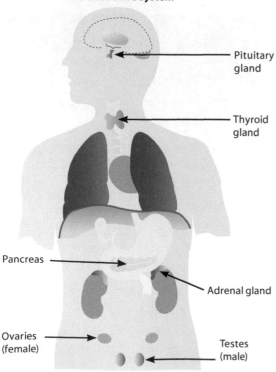

Pituitary gland

Thyroid gland

Pancreas

Adrenal gland

Ovaries (female)

Testes (male)

The pituitary gland

The pituitary is often referred to as the **master gland** because it secretes many hormones that control other processes in the body. Pituitary hormones often trigger other hormones to be released.

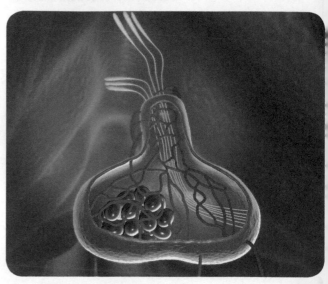

Controlling blood glucose concentration

The control system for balancing blood glucose levels involves the **pancreas**.

The pancreas monitors the blood glucose concentration and releases hormones to restore the balance. When the concentration is too high, the pancreas produces insulin that causes glucose to be absorbed from the blood by all body cells, but particularly those in the liver and muscles. These organs convert glucose to glycogen for storage until required.

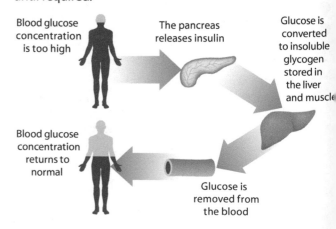

Blood glucose concentration is too high

The pancreas releases insulin

Glucose is converted to insoluble glycogen stored in the liver and muscle

Blood glucose concentration returns to normal

Glucose is removed from the blood

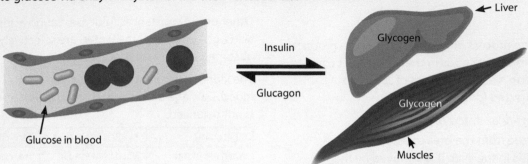

HT If blood glucose concentration is too low, the pancreas secretes glucagon. This stimulates the conversion of glycogen to glucose via enzymic systems. It is then released into the blood.

Liver

Glycogen

Insulin

Glucagon

Glycogen

Glucose in blood

Muscles

Diabetes

There are two types of diabetes.

Type I diabetes:

- is caused by the pancreas' inability to produce insulin
- results in dangerously high levels of blood glucose
- is controlled by delivery of insulin into the bloodstream via injection or a 'patch' worn on the skin
- is more likely to occur in people under 40
- is the most common type of diabetes in childhood
- is thought to be triggered by an auto-immune response where cells in the pancreas are destroyed.

Type II diabetes:

- is caused by fatty deposits preventing body cells from absorbing insulin; the pancreas tries to compensate by producing more and more insulin until it is unable to produce any more, which results in dangerously high levels of blood glucose
- is controlled by a low carbohydrate diet and exercise initially; it may require insulin in the later stages
- is more common in people over 40
- is a risk if you are obese.

SUMMARY

- **The endocrine system consists of glands that secrete hormones into the bloodstream.**
- **The pituitary gland secretes many hormones that control other processes.**
- **The pancreas monitors blood glucose concentration.**
- **Diabetes results from the body's inadequate production of insulin.**

QUESTIONS

QUICK TEST

1. Which hormone is produced in the thyroid?
2. Where is insulin produced in the body?
3. What effects does type II diabetes have on the body?

EXAM PRACTICE

1. **a)** When someone eats a meal, the pancreas responds by producing insulin.

   Describe the effects this has on the body? **[2 marks]**

   b) People who have type I diabetes have trouble balancing their blood glucose concentrations.

   How is this condition treated? **[1 mark]**

   c) How is type II diabetes caused and managed? **[2 marks]**

Water and nitrogen balance

Excretion

Excretion is the process of getting rid of waste products made by chemical reactions in the body. Don't confuse it with **egestion**, which is the loss of solid waste (mainly undigested food).

The following are excreted products.

- Urea is made from the breakdown of excess amino acids in the liver. It is removed by the kidneys along with excess water and ions and transferred to the bladder as **urine** before being released.
- **Sweat** containing water, urea and salt is excreted by sweat glands onto the surface of the skin. Sweating aids the body's cooling process.
- **Carbon dioxide** and **water** are produced by respiration and leave the body from the lungs during exhalation.

The lungs and skin don't control the loss of substances. They are simply the organs by which these substances are removed.

The kidneys

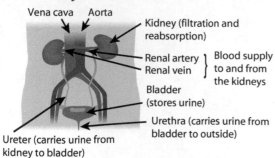

Vena cava · Aorta

Kidney (filtration and reabsorption)

Renal artery
Renal vein } Blood supply to and from the kidneys

Bladder (stores urine)

Urethra (carries urine from bladder to outside)

Ureter (carries urine from kidney to bladder)

The kidneys filter the blood, allowing urea to pass to the bladder. The filtering is carried out within the kidney by thousands of tiny **kidney tubules**.

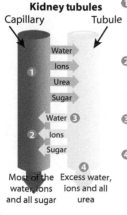

Kidney tubules
Capillary · Tubule

Water
Ions
Urea
Sugar

Water ③
Ions
Sugar

Most of the water, ions and all sugar

Excess water, ions and all urea

① **Filtration**
Lots of water plus all the small molecules are squeezed out of the blood, under pressure, into the tubules.

② **Selective reabsorption**
Useful substances, including glucose, ions and water, are reabsorbed into the blood from the tubules.

③ **Osmoregulation**
Amount of water in the blood and urine is adjusted here.

④ **Excretion of waste**
Excess water, ions and all the urea now pass to the bladder in the form of urine and are eventually released from the body.

Other useful substances (glucose, amino acids, fatty acids, glycerol and some water) are selectively re-absorbed early in the process.

The kidneys also control the balance of water in the blood. Damage occurs to red blood cells if water content is not balanced.

Ideal shape	Swollen	Shrivelled
When red blood cells (erythrocytes) are in solutions with equal concentration to their cytoplasm, they have an ideal, biconcave shape. This is because there is no net movement of water in or out.	When immersed in a solution of lower concentration (higher water concentration), the cells absorb water by osmosis. The weak cell membrane cannot resist the added water pressure and may burst.	In a more concentrated solution (lower water concentration), cells lose water by osmosis. They shrivel up and become **crenated** (have scalloped edges).

Kidney tubules (nephrons)

In terms of nephron structure, filtration, selective reabsorption and excretion of waste occur in the following regions.

Structure of the nephron

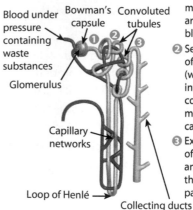

Blood under pressure containing waste substances

Bowman's capsule · Convoluted tubules

Glomerulus

Capillary networks

Loop of Henlé

Collecting ducts (lead to the ureter)

① Filtration, where all small molecules and lots of water are squeezed out of the blood and into the tubules.

② Selective reabsorption of useful substances (water, ions, glucose) back into the blood from the convoluted tubules. This may take energy in the case of glucose and ions.

③ Excretion of waste in the form of excess water, excess ions and all urea. These drain into the collecting tubules and pass to the bladder as urine.

Kidney failure

Kidneys may fail due to accidents or disease. A patient can survive with one kidney. If both kidneys are affected, two treatments are available.

Kidney transplant – involves a healthy person donating one kidney to replace two failed kidneys in another person.

Dialysis – offered to patients while they wait for the possibility of a kidney transplant. A dialysis machine removes urea and maintains levels of sodium and glucose in the blood.

This is what happens during dialysis.

1. Blood is taken from a person's vein and run into the dialysis machine, where it comes into close contact with a partially permeable membrane.
2. This separates the blood from the dialysis fluid.
3. The urea and other waste diffuse from the blood into the dialysis fluid. The useful substances remain and are transferred back to the body.

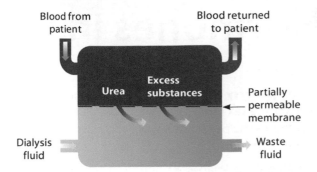

HT How urea is formed

Proteins obtained from the diet may produce a surplus of **amino acids** that need to be excreted safely.

1. First, the amino acids are deaminated in the **liver** to form ammonia.
2. Ammonia is toxic so is immediately converted to urea, which is then filtered out in the kidney.

Controlling water content

The osmotic balance of the body's fluids needs to be tightly controlled because if cells gain or lose too much water they do not function efficiently.

The amount of water re-absorbed by the kidneys is controlled by **anti-diuretic hormone (ADH)**. This is produced in the pituitary.

1. ADH directly increases the permeability of the kidney tubules to water.
2. When the water content of the blood is low (higher blood concentration), **negative feedback** operates to restore normal levels.

The effect of ADH on blood water content

Blood water level too low (salt concentration too high)	Blood water level too high (salt concentration too low)
Detected by the pituitary gland	Detected by the pituitary gland
More ADH released into the blood by pituitary gland	**Less** ADH released into the blood by pituitary gland
More water reabsorbed into the blood from the renal tubules	Less water reabsorbed into the blood from the renal tubules
Small amount of concentrated urine	Large amount of dilute urine
Normal blood water level	Normal blood water level

SUMMARY

- Urea, sweat, carbon dioxide and water are the products of excretion.
- The kidneys remove urea from the body. When kidneys fail, a patient may receive dialysis or a kidney transplant.
- The kidney carries out water balance in tandem with the hormone ADH.

QUESTIONS

QUICK TEST

1. List three substances that are selectively re-absorbed back into the bloodstream from the kidney.
2. Which organ releases ADH into the blood?

EXAM PRACTICE

1. A student is exercising.

   She produces carbon dioxide in her cells.

   a) What name is given to the process that converts substances into waste products? **[1 mark]**

   b) The student's liver produces urea.

   Which substances are changed to form urea? **[1 mark]**

Hormones in human reproduction

Puberty

During **puberty** (approximately 10–16 in girls and 12–17 in boys), the sex organs begin to produce **sex hormones**. This causes the development of secondary sexual characteristics.

In **males**, the primary sex hormone is **testosterone**.

During puberty, testosterone is produced from the testes and causes:

- production of sperm in testes
- development of muscles and penis
- deepening of the voice
- growth of pubic, facial and body hair.

In **females**, the primary sex hormone is **oestrogen**. Other sex hormones are **progesterone**, **FSH** and **LH**.

During puberty, oestrogen is produced in the ovaries and progesterone production starts when the menstrual cycle begins.

The secondary sexual characteristics are:

- ovulation and the menstrual cycle
- breast growth
- widening of hips
- growth of pubic and armpit hair.

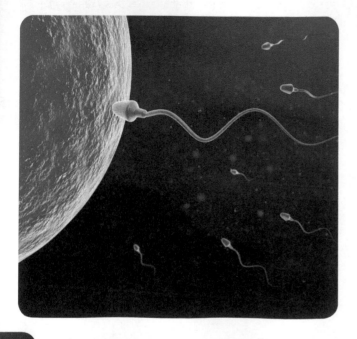

The menstrual cycle

A woman is fertile between the ages of approximately 13 and 50.

During this time, an egg is released from one of her ovaries each month and the lining of her uterus is replaced each month (approximately 28 days) to prepare for pregnancy.

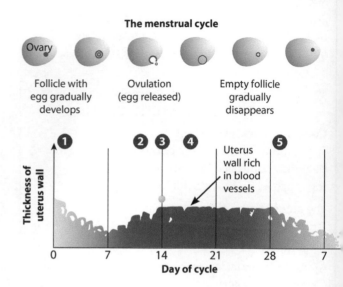

The menstrual cycle

Ovary

Follicle with egg gradually develops

Ovulation (egg released)

Empty follicle gradually disappears

Uterus wall rich in blood vessels

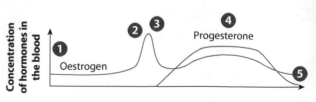

Progesterone

Oestrogen

1. Uterus lining breaks down (i.e. a period).

2. Repair of the uterus wall.
   Oestrogen causes the uterus lining to gradually thicken.

3. Egg released by the ovary.

4. Progesterone and oestrogen make the lining stay thick, waiting for a fertilised egg.

5. No fertilised egg so cycle restarts.

As well as oestrogen and progesterone, the two other hormones involved in the cycle are:

- **FSH** or follicle stimulating hormone, which causes maturation of an egg in the ovary
- **LH** or luteinising hormone, which stimulates release of an egg.

HT Negative feedback in the menstrual cycle

The four female hormones interact in a complex manner to regulate the cycle.

- **FSH** is produced in the pituitary and acts on the ovaries, causing an egg to mature. It stimulates the ovaries to produce oestrogen.

- **Oestrogen** is secreted in the ovaries and inhibits further production of FSH. It also stimulates the release of LH and promotes repair of the uterus wall after menstruation.

- **LH** is produced in the pituitary. It also stimulates release of an egg.

- **Progesterone** is secreted by the empty follicle in the ovary (left by the egg). It maintains the lining of the uterus after ovulation has occurred. It also inhibits FSH and LH.

Female reproductive system

Low progesterone levels allow FSH from the pituitary gland to stimulate the maturation of an egg (in a follicle). This in turn stimulates oestrogen production.

High levels of oestrogen stimulate a surge in LH from the pituitary gland. This triggers ovulation in the middle of the cycle.

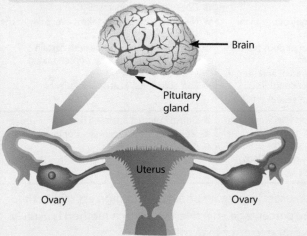

Brain

Pituitary gland

Uterus

Ovary

Ovary

SUMMARY

- **During puberty the sex organs begin to produce sex hormones.**
- **Hormones play a vital role in regulating human reproduction, especially in the female menstrual cycle.**

QUESTIONS

QUICK TEST

1. Name the four hormones involved in the menstrual cycle.

2. Name one effect of testosterone in puberty.

EXAM PRACTICE

1. **a)** State the functions of oestrogen and progesterone in the menstrual cycle. **[2 marks]**

   HT **b)** Oestrogen also stimulates release of LH.

   What is the result of this surge? **[1 mark]**

2. From the list below, circle **three** characteristics that develop as a result of puberty.

   Breast development   **Fertilisation**

   Sperm production   **Secretion of ADH**

   Menstrual cycle

   [3 marks]

Contraception and infertility

Non-hormonal contraception

Contraceptive method	Method of action	Advantages	Disadvantages
● Barrier method – condom (male + female)	Prevents the sperm from reaching the egg	82% effective ● Most effective against STIs	● Can only be used once ● May interrupt sexual activity ● Can break ● Some people are allergic to latex
● Barrier method – diaphragm	Prevents the sperm from reaching the egg	88% effective ● Can be put in place right before intercourse or 2–3 hours before ● Don't need to take out between acts of sexual intercourse	● Increases urinary tract infections ● Doesn't protect against STIs
● Intrauterine device	Prevents implantation – some release hormones	99% effective ● Very effective against pregnancy ● Doesn't need daily attention ● Comfortable ● Can be removed at any time	● Doesn't protect against STIs ● Needs to be inserted by a medical practitioner ● Higher risk of infection when first inserted ● Can have side effects such as menstrual cramping ● Can fall out and puncture the uterus (rare)
● Spermicidal agent	Kills or disables sperm	72% effective ● Cheap	● Doesn't protect against STIs ● Needs to be reapplied after one hour ● Increases urinary tract infections ● Some people are allergic to spermicidal agents
● Abstinence ● Calendar method	Refraining from sexual intercourse when an egg is likely to be in the oviduct	76% effective ● Natural ● Approved by many religions ● Woman gets to know her body and menstrual cycles	● Doesn't protect against STIs ● Calculating the ovulation period each month requires careful monitoring and instruction ● Can't have sexual intercourse for at least a week each month
● Surgical method	Vasectomy and female sterilisation	99% effective ● Very effective against pregnancy ● One-time decision providing permanent protection	● No protection against STIs ● Need to have minor surgery ● Permanent

Hormonal contraception

Contraceptive method	Method of action	Advantages	Disadvantages
● Oral contraceptive	Contains hormones that inhibit FSH production, so eggs fail to mature	91% effectiveness ● Very effective against pregnancy if used correctly ● Makes menstrual periods lighter and more regular ● Lowers risk of ovarian and uterine cancer, and other conditions ● Doesn't interrupt sexual activity	● Doesn't protect against STIs ● Need to remember to take it every day at the same time ● Can't be used by women with certain medical problems or by women taking certain medications ● Can occasionally cause side effects
● Hormone injection ● Skin patch ● Implant	Provides slow release of progesterone; this inhibits maturation and release of eggs	91–99% effectiveness depending on method used ● Lasts over many months or years ● Light or no menstrual periods ● Doesn't interrupt sexual activity	● Doesn't protect against STIs ● May require minor surgery (for implant) ● Can cause side effects

The percentage figures in the contraception tables are based on users in a whole population, regardless of whether they use the method correctly. If consistently used correctly, the percentage effectiveness of each method is usually higher. Some methods, such as the calendar method, are more prone to error than others.

HT Infertility treatment

Infertility treatment is used by couples who have problems conceiving.

Reasons for infertility

No eggs being released from the ovaries.
Endometriosis, which occurs when the tissue that lines the inside of the uterus enters other organs of the body, such as the abdomen and fallopian tubes; this reduces the maturation rate and release of eggs.
Male infertility/low sperm count.
Uterine fibroids.
Complete or partial blocking and/or scarring of the fallopian tubes.
Reduced number and quality of eggs.

Methods of treatment

Treating infertility is known as ART (assisted reproductive technology). There are a number of methods of treatment.

- Fertility drugs containing FSH and LH are given to women who do not produce enough FSH themselves. They may then become pregnant naturally.
- Clomifene therapy prevents the production of oestrogen and so inhibits negative feedback.

In vitro fertilisation (IVF) is a method in which the potential mother is given FSH and LH to stimulate the production of several eggs. Sperm is collected from the father. The sperm and eggs are then introduced together outside the body in a petri dish. One or two growing embryos can then be transplanted into the woman's uterus.

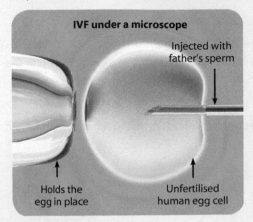

IVF under a microscope

Injected with father's sperm

Holds the egg in place

Unfertilised human egg cell

Disadvantages of IVF

It is very expensive.
It can be mentally and physically stressful.
Success rates are only approximately 40% (at the time of writing).
There is an increased risk of multiple births.

SUMMARY

- **Fertility and the possibility of pregnancy can be controlled using non-hormonal and hormonal methods of contraception.**
- **Infertility has a number of causes and may be treated using fertility drugs and surgical procedures.**

QUESTIONS

QUICK TEST

1. Which non-hormonal contraceptive is the least effective? What is the benefit of this method?

2. Why are condoms effective against the spread of HIV?

EXAM PRACTICE

1. State one advantage and one disadvantage of using a condom barrier method of contraception. **[2 marks]**

HT 2. IVF is one method of treating infertility.

   Explain how this is carried out. **[3 marks]**

Sexual and asexual reproduction

One of the basic characteristics of life is **reproduction**. This is the means by which a species continues. If sufficient offspring are not produced, the species becomes **extinct**.

Sexual reproduction

Sexual reproduction is where a male **gamete** (e.g. sperm) meets a female gamete (e.g. egg). This fusion of the two gametes is called fertilisation and may be internal or external.

Gametes are produced by **meiosis** in the sex organs.

Asexual reproduction

Asexual reproduction does not require different male and female cells. Instead, genetically identical clones are produced from mitosis. These may just be individual cells, as in the case of yeast, or whole multicellular organisms, e.g. aphids.

Many organisms can reproduce using both methods, depending on the environmental conditions.

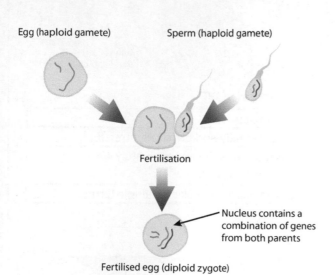

Egg (haploid gamete)

Sperm (haploid gamete)

Fertilisation

Nucleus contains a combination of genes from both parents

Fertilised egg (diploid zygote)

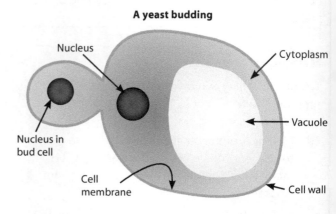

A yeast budding

Nucleus

Cytoplasm

Nucleus in bud cell

Vacuole

Cell membrane

Cell wall

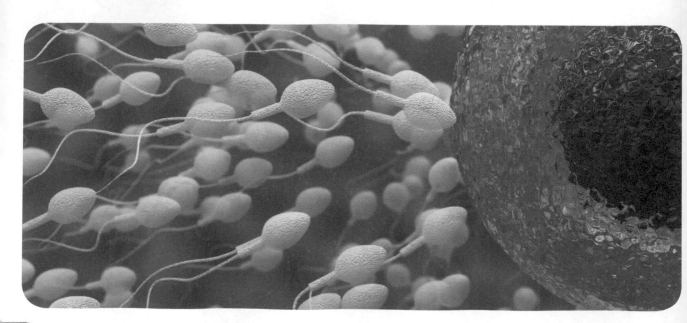

Comparing sexual and asexual reproduction

Advantages of sexual reproduction
• Produces **variation** in offspring through the process of meiosis, where genes are 'shuffled.' Variation is increased by **random fusion of gametes**.
• Survival advantage gained when the environment changes because different genetic types have more chance of producing well-adapted offspring.
• Humans can make use of sexual reproduction through **selective breeding**. This enhances food production.

Disadvantages of sexual reproduction
• Relatively slow process.
• Variation can be a disadvantage in stable environments.
• More resources required than for asexual reproduction, e.g. energy, time.
• Results of selective breeding are unpredictable and might lead to genetic abnormalities from 'in-breeding'.

Advantages of asexual reproduction
• Only one parent required.
• Fewer resources (energy and time) need to be devoted to finding a mate.
• Faster than sexual reproduction – survival advantage of producing many offspring in a short period of time.
• Many identical offspring of a well-adapted individual can be produced to take advantage of favourable conditions.

Disadvantages of asexual reproduction
• Offspring may not be well adapted in a changing environment.

Asexual reproduction

Sexual reproduction

SUMMARY

● One of the basic characteristics of life is reproduction. This is the means by which a species continues. If sufficient offspring are not produced, the species becomes extinct.

● Reproduction may be sexual or asexual.

● Some organisms reproduce using both means.

QUESTIONS

QUICK TEST

1. What name is given to the fusion of two gametes?

2. What happens to a species if not enough offspring are produced?

3. Give one advantage of asexual reproduction.

4. What is external fertilisation?

EXAM PRACTICE

1. State one advantage and one disadvantage of reproducing by sexual means. **[2 marks]**

2. a) The malarial parasite can live in two different organisms.

   Suggest a survival advantage in this. **[1 mark]**

   b) Name one other type of organism that combines asexual and sexual reproduction. **[1 mark]**

DNA

DNA and the genome

The nucleus of each cell contains a complete set of genetic instructions called the **genetic code**. The information is carried as genes, which are small sections of DNA found on **chromosomes**. The genetic code controls cell activity and, consequently, characteristics of the whole organism.

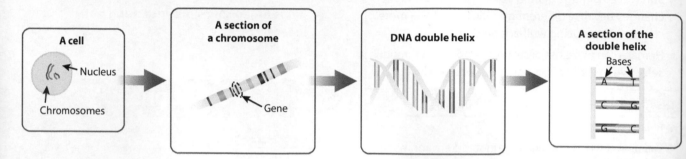

A cell — Nucleus, Chromosomes

A section of a chromosome — Gene

DNA double helix

A section of the double helix — Bases, A, T, G, C, G, C

DNA facts

- DNA is a polymer.

- It is made of two strands coiled around each other called a **double helix**.

- The genetic code is in the form of nitrogenous **bases**.

- Bases bond together in pairs forming hydrogen bond cross-links.

- The structure of DNA was discovered in 1953 by **James Watson** and **Francis Crick**, using experimental data from **Rosalind Franklin** and **Maurice Wilkins**.

- A single gene codes for a particular sequence of **amino acids**, which, in turn, make up a single **protein**.

The human genome

The **genome** of an organism is the entire genetic material present in an adult cell of an organism.

The Human Genome Project (HGP)

The HGP was an international study. Its purpose was to map the complete set of genes in the human body.

HGP scientists worked out the code of the human genome in three ways. They:

- determined the sequence of all the bases in the genome

- drew up maps showing the locations of the genes on chromosomes

- produced linkage maps that could be used for tracking inherited traits from generation to generation, e.g. for genetic diseases. This could then lead to targeted treatments for these conditions.

The results of the project, which involved collaboration between UK and US scientists, were published in 2003. Three billion base pairs were determined.

The mapping of the human genome has enabled anthropologists to work out historical human migration patterns. This has been achieved by collecting and analysing DNA samples from many people across the globe. The study is called the **Genographic Project**.

15 000   4500

25 000

40 000

100 000   70 000

12 000

200 000   1500

30 000

50 000

1500

Homo sapiens
Homo neanderthalensis
Homo erectus

DNA structure and base sequences

The four bases in DNA are A, T, C and G. The code is 'read' on one strand of DNA. Three consecutive bases (a **triplet**) code for one particular amino acid. The sequence of these triplets determines the structure of a whole protein.

The bases are attached to a sugar phosphate **backbone**. These form a basic unit called a nucleotide.

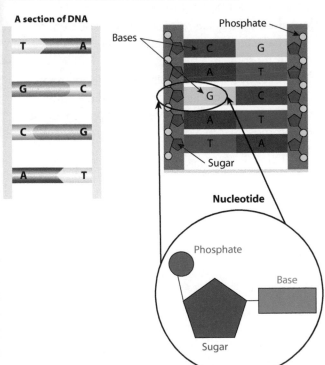

A section of DNA

T — A
G — C
C — G
A — T

Bases

Phosphate

C   G
A   T
G   C
A   T
T   A

Sugar

Nucleotide

Phosphate

Base

Sugar

SUMMARY

- Each cell in the body contains a nucleus with chromosomes.
- Genes are small sections of DNA found on chromosomes. The four bases in DNA are A, T, C and G.
- The Human Genome Project has produced a complete map of all the genes in the body.

QUESTIONS

QUICK TEST

1. Explain how the Human Genome Project has advanced medical science.

2. What are the four bases in DNA?

EXAM PRACTICE

1. Explain how scientists have used knowledge about the human genome to contribute to the **Genographic project**.   [2 marks]

2. Describe how a nucleotide from the DNA molecule is made up.   [3 marks]

The genetic code

(HT) Mutations

Mutations (genetic variants):

● are changes to the structure of a DNA molecule

● occur continuously during the cell division process or as a result of external influences, e.g. exposure to radioactive materials or emissions such as X-rays or UV light

● usually have a neutral effect as amino acids may still be produced or the proteins produced work in the same way

● may result in harmful conditions or, more rarely, beneficial traits

● result in a change in base sequence and therefore changes in the amino acid sequence and protein structure.

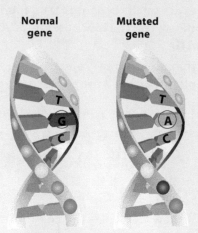

| Normal gene | Mutated gene |

The G base is substituted for an A base

Proteins produced as a result of mutation may no longer be able to carry out their function. This is because they have a different 3D structure. For example, an enzyme's active site may no longer fit with its substrate, or a structural protein may lose its strength.

Changes in the base sequence may be passed on to daughter cells when cell division occurs. This in turn may lead to offspring having genetic conditions.

(WS) During your course, you will be expected to recognise, draw and interpret scientific diagrams.

The way complementary strands of DNA are arranged can be worked out once you know that base T bonds with A and base C bonds with G.

Can you write out the complementary (DNA) strand to this sequence?

_____

| A | T | T | A | C | G | T | G | A | G | C | C |

Monohybrid crosses

Most characteristics or **traits** are the result of multiple alleles interacting but some are controlled by a single gene. Examples include fur colour in mice and the shape of earlobes in humans.

These genes exist as pairs called alleles on homologous chromosomes.

Alleles are described as **dominant** or **recessive**.

● A **dominant** allele controls the development of a characteristic even if it is present on only one chromosome in a pair.

● A **recessive** allele controls the development of a characteristic only if a dominant allele is not present, i.e. if the recessive allele is present on both chromosomes in a pair.

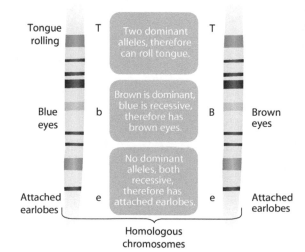

Tongue rolling	T	Two dominant alleles, therefore can roll tongue.	T	
Blue eyes	b	Brown is dominant, blue is recessive, therefore has brown eyes.	B	Brown eyes
Attached earlobes	e	No dominant alleles, both recessive, therefore has attached earlobes.	e	Attached earlobes

Homologous chromosomes

If both chromosomes in a pair contain the same allele of a gene, the individual is described as being **homozygous** for that gene or condition.

If the chromosomes in a pair contain different alleles of a gene, the individual is **heterozygous** for that gene or condition.

The combination of alleles for a particular characteristic is called the **genotype**. For example, the genotype for a homozygous dominant tongue-roller would be TT. The fact that this individual is able to roll their tongue is termed their **phenotype**.

Other examples are:

● bb (genotype), blue eyes (phenotype)
● EE or Ee (genotype), unattached/pendulous earlobes (phenotype).

When a characteristic is determined by just one pair of alleles, as with eye colour and tongue rolling, it is called **monohybrid inheritance**.

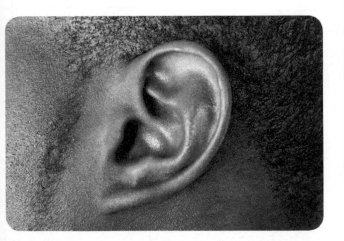

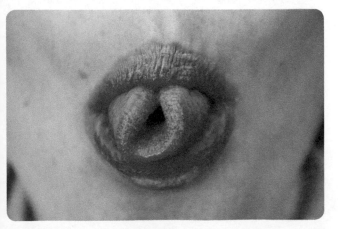

SUMMARY

● **Bases in DNA always bond T to A and G to C.**
● **HT** **Mutations are genetic variants and may result in harmful conditions or occasionally beneficial traits.**
● **Alleles can be dominant or recessive.**

QUESTIONS

QUICK TEST

1. What are alleles?

2. What is the difference between a dominant and a recessive allele?

3. What are the four bases of DNA?

EXAM PRACTICE

1. Explain how a recessive allele could control the development of a characteristic. **[2 marks]**

HT 2. Explain how a mutation might result in a different protein being produced by a cell. **[3 marks]**

Inheritance and genetic disorders

Genetic diagrams

Genetic diagrams are used to show all the possible combinations of alleles and outcomes for a particular gene. They use:

- capital letters for dominant alleles
- lower-case letters for recessive alleles.

For eye colour, brown is dominant and blue is recessive. So B represents a brown allele and b represents a blue allele.

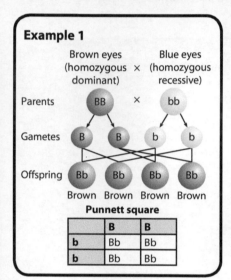

Example 1

Brown eyes (homozygous dominant) × Blue eyes (homozygous recessive)

Parents: BB × bb
Gametes: B B b b
Offspring: Bb Bb Bb Bb
Brown Brown Brown Brown

Punnett square

	B	B
b	Bb	Bb
b	Bb	Bb

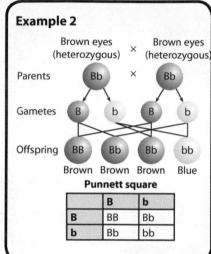

Example 2

Brown eyes (heterozygous) × Brown eyes (heterozygous)

Parents: Bb × Bb
Gametes: B b B b
Offspring: BB Bb Bb bb
Brown Brown Brown Blue

Punnett square

	B	b
B	BB	Bb
b	Bb	bb

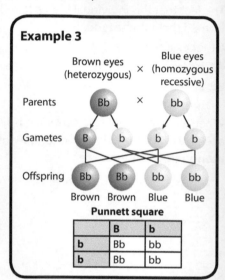

Example 3

Brown eyes (heterozygous) × Blue eyes (homozygous recessive)

Parents: Bb × bb
Gametes: B b b b
Offspring: Bb Bb bb bb
Brown Brown Blue Blue

Punnett square

	B	b
b	Bb	bb
b	Bb	bb

WS You need to be able to interpret genetic diagrams and work out ratios of offspring.

- In Example 2 above, the phenotypes of the offspring are 'brown eyes' and 'blue eyes'. As there are potentially three times as many brown-eyed children as blue-eyed, the phenotypes are said to be in a 3:1 ratio.

- In Example 3 above, the ratio would be 1:1 because half of the theoretical offspring are brown-eyed and half blue-eyed. Another way of saying this is that the probability of parents producing a brown-eyed child is 50%, or ½, or 0.5.

Most traits result not from one pair of alleles but from multiple genes interacting, e.g. inheritance of blood groups in the **ABO** system.

HT In exams, you may be asked to construct your own punnett squares to solve genetic cross problems like the ones above.

Family trees

Family trees are another way of showing how genetic traits can be passed on. Here is an example.

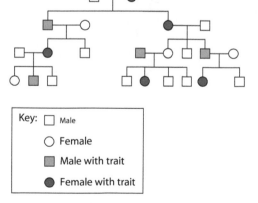

Key:
- ☐ Male
- ○ Female
- ■ Male with trait
- ● Female with trait

Inheritance of sex

Sex in humans/mammals is determined by whole chromosomes. These are the 23rd pair and are called sex chromosomes. There is an 'X' chromosome and a smaller 'Y' chromosome. The other 22 chromosome pairs carry the remainder of genes coding for the rest of the body's characteristics.

All egg cells carry X chromosomes. Half the sperm carry X chromosomes and half carry Y chromosomes. The sex of an individual depends on whether the egg is fertilised by an X-carrying sperm or a Y-carrying sperm.

If an X sperm fertilises the egg it will become a girl. If a Y sperm fertilises the egg it will become a boy. The chances of these events are equal, which results in approximately equal numbers of male and female offspring.

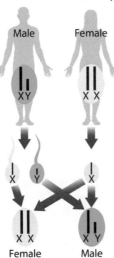

Inherited diseases

Some disorders are caused by a 'faulty' gene, which means they can be **inherited**. One example is **polydactyly**, which is caused by a dominant allele and results in extra fingers or toes. The condition is not life-threatening.

Cystic fibrosis, on the other hand, can limit life expectancy. It causes the mucus in respiratory passages and the gut lining to be very thick, leading to build-up of phlegm and difficulty in producing correct digestive enzymes.

Cystic fibrosis is caused by a recessive allele. This means that an individual will only exhibit symptoms if both recessive alleles are present in the genotype. Those carrying just one allele will not show symptoms, but could potentially pass the condition on to offspring. Such people are called **carriers**.

Conditions such as cystic fibrosis are mostly caused by **faulty alleles** that are **recessive**.

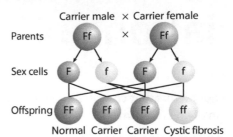

Knowing that there is a 1 in 4 chance that their child might have cystic fibrosis gives parents the opportunity to make decisions about whether to take the risk and have a child. This is a very difficult decision to make.

Technology has advanced and it is now possible to screen embryos for genetic disorders.

● If an embryo has a life-threatening condition it could be destroyed, or simply not be implanted if IVF was applied.

● Alternatively, new gene therapy techniques might be able to reverse the effects of the condition, resulting in a healthy baby.

As yet, these possibilities have to gain approval from ethics committees – some people think that this type of 'interference with nature' could have harmful consequences.

SUMMARY

● Genetic diagrams show all possible combinations of alleles and outcomes for a gene. Family trees are another way of showing how genetic traits are passed on.

QUESTIONS

QUICK TEST

1. If a homozygous brown-eyed individual is crossed with a homozygous blue-eyed individual, what is the probability of them producing a blue-eyed child?

EXAM PRACTICE

1. A man who knows he is a carrier of cystic fibrosis is thinking of having children with his partner. His partner does not suffer from cystic fibrosis, nor is she a carrier.

   Suggest what a genetic counsellor might advise the couple.

   Use genetic diagrams to aid your explanation and state any probabilities of offspring produced. **[6 marks]**

Variation and evolution

Variation

The two major factors that contribute to the appearance and function of an organism are:

- **genetic information** passed on from parents to offspring

- **environment** – the conditions that affect that organism during its lifetime, e.g. climate, diet, etc.

These two factors account for the large variation we see **within** and **between** species. In most cases, both of these factors play a part.

Evolution

Put simply, evolution is the theory that all organisms have arisen from simpler life forms over billions of years. It is driven by the **mechanism** of **natural selection**. For natural selection to occur, there must be genetic variation between individuals of the same species. This is caused by mutation or new combinations of genes resulting from sexual reproduction.

Most mutations have no effect on the phenotype of an organism. Where they do, and if the environment is changing, this can lead to relatively rapid change.

Evidence for evolution

Evidence for evolution comes from many sources. It includes:

- comparing genomes of different organisms

- studying embryos and their similarities

- looking at changes in species during modern times, e.g. antibiotic resistance in bacteria

- comparing the anatomy of different forms

- the fossil record.

One of the earliest sources of evidence for evolution was the discovery of fossils.

Further evidence for evolution – the pentadactyl limb

If you compare the forelimbs of a variety of vertebrates, you can see that the bone structures are all variations on a five-digit form, whether it be a leg, a wing or a flipper. This suggests that there was a common ancestral form from which these organisms developed.

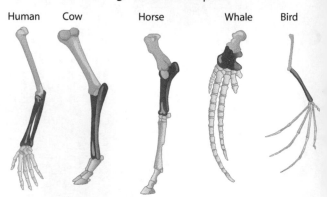

Human Cow   Horse   Whale   Bird

How fossils are formed

1 When an animal or plant dies, the processes of decay usually cause all the body tissues to break down. In rare circumstances, the organism's body is rapidly covered and oxygen is prevented from reaching it. Instead of decay, fossilisation occurs.

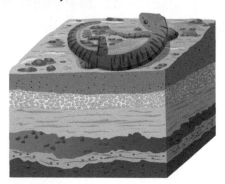

2 Over hundreds of thousands of years, further sediments are laid down and compress the organism's remains.

3 Parts of the organism, such as bones and teeth, are **replaced by minerals** from solutions in the rock.

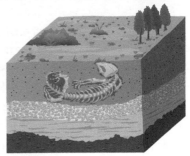

4 Earth upheavals, e.g. **tectonic plate movement**, bring sediments containing the fossils nearer the Earth's surface.

5 Erosion of the rock by wind, rain and rivers exposes the fossil. At this stage, the remains might be found and excavated by **paleontologists**.

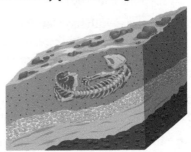

Fossils can also be formed from footprints, burrows and traces of tree roots.

By comparing different fossils and where they are found in the rock layers, paleontologists can gain insights into how one form may have developed into another.

Difficulties occur with earlier life forms because many were **soft-bodied** and therefore not as well-preserved as organisms with bones or shells. Any that are formed are easily destroyed by Earth movements. As a result of this, scientists cannot be certain about exactly how life began.

Extinction

The fossil record provides evidence that most organisms that once existed have become **extinct**. In fact, there have been at least five **mass** extinctions in geological history where most organisms died out. One of these coincides with the disappearance of dinosaurs.

Causes of extinction include:

● **catastrophic events**, e.g. volcanic eruptions, asteroid collisions

● changes to the environment over geological time

● new **predators**

● new **diseases**

● new, more successful **competitors**.

SUMMARY

● **Genetic information, and environment are the two major factors that contribute to variation.**

● **Evolution is the theory that all organisms have developed from simple life forms over billions of years.**

● **The fossil record tells us that most organisms that once existed have become extinct.**

● **Speciation occurs where members of an original population can no longer interbreed.**

QUESTIONS

QUICK TEST

1. Suggest two pieces of evidence that support the theory of evolution through natural selection.

2. Suggest two causes of extinction.

EXAM PRACTICE

1. The fore-limbs found in human, cow, horse, whale and bird are all based on the pentadactyl limb which means a five-digited limb.

   Scientists believe that the whale may have evolved from a horse-like ancestor that lived in swampy regions millions of years ago.

   Suggest how whales could have evolved from a horse-like mammal. In your answer, use Darwin's theory of natural selection. **[5 marks]**

Darwin and evolution

Human evolution

Modern man has evolved from a common ancestor that gave rise to all the primates: gorillas, chimpanzees and orangutans. DNA comparisons have shown we are most closely related to the chimpanzee.

The evolution of man can be traced back over the last four to five million years. Over this period of time, the **human form** (hominid) has developed:

- a more upright, bipedal stance
- less body hair
- a smaller and less 'domed' forehead
- greater intelligence and use of tools, initially from stone. These tools can be dated using scientific techniques, e.g. radiometric dating and carbon dating.

There have been significant finds of fossils that give clues to human evolution.

1. **Ardi** – at 4.4 million years old, this is the oldest, most complete hominid skeleton.
2. **Lucy** – from 3.2 million years ago, this is one of the first fossils to show an upright walking stance.
3. **Proconsul skull** – discovered by **Mary Leakey** and her husband; the hominid is thought to be from about 1.6 million years ago.

Darwin's theory of evolution through natural selection

Within a population of organisms there is a range of variation among individuals. This is caused by genes. Some differences will be beneficial; some will not.

Beneficial characteristics make an organism more likely to survive and pass on their genes to the next generation. This is especially true if the environment is changing. This ability to be successful is called **survival of the fittest**.

Species that are not well adapted to their environment may become extinct. This process of change is summed up in the theory of evolution through **natural selection**, put forward by **Charles Darwin** in the nineteenth century.

Many theories have tried to explain how life might have come about in its present form.

However, Darwin's theory is accepted by most scientists today. This is because it explains a wide range of observations and has been discussed and tested by many scientists.

Darwin's theory can be reduced to five ideas.

They are:
- variation
- competition
- survival of the fittest
- inheritance
- extinction.

Darwin's ideas are illustrated in the following two examples.

Example 1: peppered moths

Variation – most peppered moths are pale and speckled. They are easily camouflaged amongst the lichens on silver birch tree bark. There are some rare, dark-coloured varieties (that originally arose from genetic mutation). They are easily seen and eaten by birds.

Competition – in areas with high levels of air pollution, lichens die and the bark becomes discoloured by soot. The lighter peppered moths are now put at a competitive disadvantage.

Survival of the fittest – the dark (melanic) moths are now more likely to avoid detection by pedators.

Inheritance – the genes for dark colour are passed on to offspring and gradually become more common in the general population.

Extinction – if the environment remains polluted, the lighter form is more likely to become extinct.

Dark peppered moth
Peppered moth

Example 2: methicillin-resistant bacteria

The resistance of some bacteria to antibiotics is an increasing problem. MRSA bacteria have become more common in hospital wards and are difficult to eradicate.

Variation – bacteria mutate by chance, giving them a resistance to antibiotics.

Competition – the non-resistant bacteria are more likely to be killed by the antibiotic and become less competitive.

Survival of the fittest – the antibiotic-resistant bacteria survive and reproduce more often.

Inheritance – resistant bacteria pass on their genes to a new generation; the gene becomes more common in the general population.

Extinction – non-resistant bacteria are replaced by the newer, resistant strain.

To slow down the rate at which new, resistant strains of bacteria can develop:

- doctors are urged not to prescribe antibiotics for obvious viral infections or for mild bacterial infections
- patients should complete the full course of antibiotics to ensure that **all** bacteria are destroyed (so that none will survive to mutate into resistant strains).

Species become more and more specialised as they evolve and adapt to their environmental conditions.

The point at which a new species is formed occurs when the original population can no longer interbreed with the newer, 'mutant' population. For this to occur, **isolation** needs to happen.

Speciation

- Groups of the same species that are separated from each other by physical boundaries (like mountains or seas) will not be able to breed and share their genes. This is called **geographical isolation**.
- Over long periods of time, separate groups may specialise so much that they cannot successfully breed any longer and so two new species are formed – this is **reproductive isolation**.

SUMMARY

- **The ability to be successful is called survival of the fittest. Species that are not well adapted to their environment may become extinct.**
- **This process is the basis of the theory of evolution by natural selection, put forward by Charles Darwin.**

QUESTIONS

QUICK TEST

1. Define evolution.
2. What is the name of the oldest most complete hominid?
3. State two examples where natural selection has been observed by scientists in recent times.

EXAM PRACTICE

1. Alfred Wallace lived at the same time as Charles Darwin and also wrote papers on the idea of natural selection. In 1858 he wrote the following:

   "I ask myself: how and why do species change, and why do they change into new and well-defined species. Why and how do they become so exactly adapted to distinct modes of life; and why do all the intermediate grades die out?"

   Wallace had no knowledge of genes or DNA structure.

   Knowing what we know now, how would you answer his question about how organisms become 'exactly adapted?' **[2 marks]**

Selective breeding and genetic engineering

Selective breeding

Farmers and dog breeders have used the principles of selective breeding for thousands of years by keeping the best animals and plants for breeding.

For example, to breed Dalmatian dogs, the spottiest dogs have been bred through the generations to eventually get Dalmatians. The factor most affected by selective breeding in dogs is probably temperament. Most breeds are either naturally obedient to humans or are trained to be so.

This is the process of selective breeding.

Select the desired characteristics in parents. → Allow the individuals to breed (or cross-pollinate if you are dealing with plants). → Select the desired offspring and allow them to become parents of the next generation.

This process has to be repeated many times to get the desired results.

Advantages of selective breeding
● It results in an organism with the 'right' characteristics for a particular function.
● In farming and horticulture, it is a more efficient and economically viable process than natural selection.

Disadvantages of selective breeding
● Intensive selective breeding reduces the gene pool – the range of alleles in the population decreases so there is **less variation**.
● Lower variation reduces a species' ability to respond to environmental change.
● It can lead to an accumulation of harmful recessive characteristics (in-breeding), e.g. restriction of respiratory pathways and dislocatable joints in bulldogs.

Examples of selective breeding

Modern food plants

Three of our modern vegetables have come from a single ancestor by selective breeding. (Remember, it can take many, many generations to get the desired results.)

Selective breeding in plants has also been undertaken to produce:

● disease resistance in crops

● large, unusual flowers in garden plants.

Modern cattle

Selective breeding can contribute to improved yields in cattle. Here are some examples.

● **Quantity of milk** – years of selecting and breeding cattle that produce larger than average quantities of milk has produced herds of cows that produce high daily volumes of milk.

● **Quality of milk** – as a result of selective breeding, Jersey cows produce milk that is rich and creamy, and can therefore be sold at a higher price.

● **Beef production** – the characteristics of the Hereford and Angus varieties have been selected for beef production over the past 200 years or more. They include hardiness, early maturity, high numbers of offspring and the swift, efficient conversion of grass into body mass (meat).

Genetic engineering

All living organisms use the same basic genetic code (DNA). So genes can be transferred from one organism to another in order to deliberately change the recipient's characteristics. This process is called genetic engineering or genetic modification (GM).

Altering the genetic make-up of an organism can be done for many reasons.

● **To improve resistance to herbicides**: for example, soya plants are genetically modified by inserting a gene that makes them resistant to a herbicide. When the crop fields are sprayed with the herbicide only the weeds die, leaving the soya plants without competition so they can grow better. Resistance to frost or disease can also be genetically engineered. Bigger yields result.

● **To improve the quality of food**: for example, bigger and more tasty fruit.

● **To produce a substance you require**: for example, the gene for human insulin can be inserted into bacteria or fungi, to make human insulin on a large scale to treat diabetes.

● **Disease resistance**: crop plants receive genes that give them resistance to the bacterium *Bacillus thuringiensis*.

Advantages of genetic engineering

- It allows organisms with new features to be produced rapidly.
- It can be used to make biochemical processes cheaper and more efficient.
- In the future, it may be possible to use genetic engineering to change a person's genes and cure certain disorders, e.g. cystic fibrosis. This is an area of research called gene therapy.

Disadvantages of genetic engineering

- Transplanted genes may have unexpected harmful effects on human health.
- Some people are worried that GM plants may cross-breed with wild plants and release their new genes into the environment.

HT Producing insulin

The following method is used to produce insulin.

1. The human gene for insulin production is identified and removed using a restriction enzyme, which cuts through the DNA strands in precise places.

2. The same restriction enzyme is used to cut open a ring of bacterial vector DNA (a plasmid).

3. Other enzymes called ligases are then used to insert the section of human DNA into the plasmid. The DNA can be 'spliced' in this way because the ends of the strands are 'sticky'.

4. The plasmid is reinserted into a bacterium, which starts to divide rapidly. As it divides, it replicates the plasmid.

5. The bacteria are cultivated on a large scale in fermenters. Each bacterium carries the instructions to make insulin. When the bacteria make the protein, commercial quantities of insulin are produced.

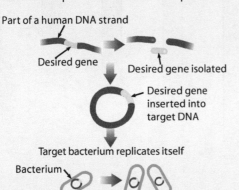

Part of a human DNA strand

Desired gene

Desired gene isolated

Desired gene inserted into target DNA

Target bacterium replicates itself

Bacterium

Sometimes, other vectors are used to introduce human DNA into organisms, e.g. viruses. It is important that the hybrid genes are transferred to the host organism at an early stage of its development, e.g. the egg or the embryo stage. As the cells are quite undifferentiated, the desired characteristics from the inserted DNA are more likely to develop.

SUMMARY

- **Selective breeding is done to ensure desirable characteristics are passed on.**

- **Genetic modification is when genes are transferred from one organism to another to change the recipient's characteristics.**

QUESTIONS

QUICK TEST

1. What advantages does genetic engineering have over selective breeding?

HT 2. What is a plasmid?

EXAM PRACTICE

1. State one advantage and one disadvantage of selective breeding. **[2 marks]**

HT 2. Describe the process where artificial insulin is produced via genetic engineering.

   State any enzymes involved and how the final product is obtained. The first stage is completed for you. **[5 marks]**

   Stage 1: Human gene for insulin identified

   Stage 2: ....

   Stage 3: ....

   Stage 4: ....

   Stage 5: ....

   Stage 6: ....

Classification

The origins of classification

In the past, observable characteristics were used to place organisms into categories.

In the eighteenth century, **Carl Linnaeus** produced the first classification system. He developed a hierarchical arrangement where larger groups were subdivided into smaller ones.

Kingdom	Largest group
Phylum	
Class	
Order	
Family	
Genus	
Species	Smallest group

Linnaeus also developed a binomial system for naming organisms according to their genus and species. For example, the common domestic cat is *Felis catus*. Its full classification would be:

● Kingdom: *Animalia*

● Phylum: *Chordata*

● Class: *Mammalia*

● Order: *Carnivora*

● Family: *Felidae*

● Genus: *Felis*

● Species: *Catus*.

Linnaeus' system was built on and resulted in a **five-kingdom system**. Developments that contributed to the introduction of this system included improvements in microscopes and a more thorough understanding of the biochemical processes that occur in all living things. For example, the presence of particular chemical pathways in a range of organisms indicated that they probably had a common ancestor and so were more closely related than organisms that didn't share these pathways.

Kingdom	Features	Examples
Plants	Cellulose cell wall Use light energy to produce food	Flowering plants Trees Ferns Mosses
Animals	Multicellular Feed on other organisms	Vertebrates Invertebrates
Fungi	Cell wall of chitin Produce spores	Toadstools Mushrooms Yeasts Moulds
Protoctista Protozoa	Mostly single-celled organisms	Amoeba Paramecium
Prokaryotes	No nucleus	Bacteria Blue–green algae

The classification diagram below illustrates how different lines of evidence can be used. The classes of vertebrates share a common ancestor and so are quite closely related. Evidence for this lies in comparative anatomy (e.g. the pentadactyl limb) and similarities in biochemical pathways.

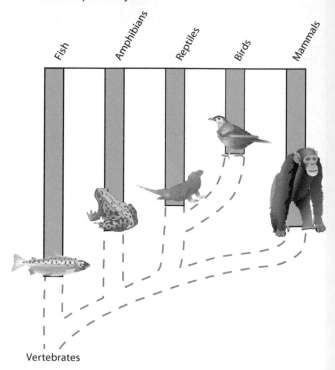

Vertebrates

In more recent times, improvements in science have led to a **three-domain system** developed by **Carl Woese**. In this system organisms are split into:

● **archaea** (primitive bacteria)

- **bacteria** (true bacteria)
- **eukaryota** (including Protista, fungi, plants and animals).

Further improvements in science include chemical analysis and further refinements in comparisons between non-coding sections of DNA.

Evolutionary trees

Tree diagrams are useful for depicting relationships between similar groups of organisms and determining how they may have developed from common ancestors. Fossil evidence can be invaluable in establishing these relationships.

Here, two species are shown to have evolved from a common ancestor.

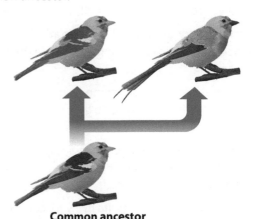

Common ancestor

WS You need to understand how new evidence and data leads to changes in models and theories.

In the case of the three-domain system, a more accurate and cohesive classification structure was proposed as a result of improvements in microscopy and increasing knowledge of organisms' internal structures.

These apes share a common ancestor

SUMMARY

- There is a huge variety of living organisms. Scientists group or classify them using shared characteristics.
- This is important because it helps to: work out how organisms evolved on Earth; understand how organisms coexist in ecological communities and identify and monitor rare organisms that are at risk from extinction.

QUESTIONS

QUICK TEST

1. What do the first and second words in the binomial name of an organism mean?

2. Who developed the recent three-domain system?

EXAM PRACTICE

1. In the Linnean system of classification, organisms are placed within a hierarchy of organisation.

   a) The following table shows this hierarchy with some missing levels. Complete the missing parts.

Kingdom
....
Class
....
Family
....
Species

   [3 marks]

   b) Linnaeus also developed the bionomial system of classification. The domestic dog has the following classification:

   Kingdom – *Animalia*; Class – *Mammalia*; Family – *Canidae*; Genus – *Canis*; Species – *familiaris*

   Use this information to suggest the binomial name of the domestic dog. **[1 mark]**

Organisms and ecosystems

Communities

Ecosystems are physical environments with a particular set of conditions (**abiotic** factors), plus all the organisms that live in them. The organisms interact through competition and predation. An ecosystem can support itself without any influx of other factors or materials. Its energy source (usually the Sun) is the only external factor.

Other terms help to describe aspects of the environment.

- The **habitat** of an animal or plant is the part of the physical environment where it lives. There are many types of habitat, each with particular characteristics, e.g. pond, hedgerow, coral reef.

- A **population** is the number of individuals of a species in a defined area.

- A **community** is the total number of individuals of all the different populations of organisms that live together in a habitat at any one time.

An organism must be well-suited to its habitat to be able to compete with other species for limited environmental resources. Even organisms within the same species may compete in order to survive and breed. Organisms that are specialised in this way are restricted to that type of habitat because their adaptations are unsuitable elsewhere.

Resources that plants compete over include:
- light
- space
- water
- minerals.

Animals compete over:
- food
- mates.
- territory.

Interdependence

In communities, each species may depend on other species for food, shelter, pollination and seed dispersal. If one species is removed, it may have knock-on effects for other species.

Stable communities contain species whose numbers fluctuate very little over time. The populations are in balance with the physical factors that exist in that habitat. Stable communities include **tropical rainforests** and ancient oak **woodlands**.

Adaptations

Adaptations:

- are special features or behaviours that make an organism particularly well-suited to its environment and better able to compete with other organisms for limited resources

- can be thought of as a biological solution to an environmental challenge – evolution provides the solution and makes species fit their environment.

Animals have developed in many different ways to become well adapted to their environment and to help them survive. Adaptations are usually of three types:

- **Structural** – for example, skin colouration in chameleons provides camouflage to hide them from predators.

- **Functional** – for example, some worms have blood with a high affinity for oxygen; this helps them to survive in anaerobic environments.

- **Behavioural** – for example, penguins huddle together to conserve body heat in the Antarctic habitat.

Look at the **polar bear** and its life in a very cold climate. It has:

- small ears and large bulk to reduce its surface area to volume ratio and so reduce heat loss
- a large amount of insulating fat (blubber)
- thick white fur for insulation and camouflage
- large feet to spread its weight on snow and ice
- fur on the soles of its paws for insulation and grip
- powerful legs so it is a good swimmer and runner, which enables it to catch its food
- sharp claws and teeth to capture prey.

The **cactus** is well adapted to living in a desert habitat. It:

- has a rounded shape, which gives a small surface area to volume ratio and therefore reduces water loss
- has a thick waxy cuticle to reduce water loss
- stores water in a spongy layer inside its stem to resist drought
- has sunken stomata, meaning that air movement is reduced, minimising loss of water vapour through them
- has leaves that take the form of spines to reduce water loss and to protect the cactus from predators.

Some organisms have biochemical adaptations. **Extremophiles** can survive extreme environmental conditions. For example:

- bacteria living in deep sea vents have optimum temperatures for enzymes that are much higher than 37°C
- icefish have antifreeze chemicals in their bodies, which lower the freezing point of body fluids
- some organisms can resist high salt concentrations or pressure.

WS During your course you will be asked to suggest explanations for observations made in the field or laboratory. These include:

- suggesting factors for which organisms are competing in a certain habitat
- giving possible adaptations for organisms in a habitat.

For example, low-lying plants in forest ecosystems often have specific adaptations for maximising light absorption as they are shaded by taller plants. Adaptations might include leaves with a large surface area and higher concentrations of photosynthetic pigments to absorb the correct wavelengths and lower intensities of light.

SUMMARY

- In a community, each species may depend on other species for food, shelter, pollination and seed dispersal. They are interdependent.
- Organisms need to be well adapted within the ecosystem in order to survive.
- Adaptations may be structural, functional or behavioural.
- Extremophiles can survive in extreme environments, e.g. very low temperatures.

QUESTIONS

QUICK TEST

1. What is an ecosystem?
2. What three resources do animals compete over?
3. Give two examples of extremophiles.

EXAM PRACTICE

1. Amazonian tree frogs have adaptations suited to living in equatorial rainforests.

   Explain how each of the adaptations listed below helps the frog to survive.

   i) Sticky pads on its 'fingers' and toes.

   ii) Hides its bright colours when asleep by closing its eyes and tucking its feet under its body.

   iii) Long sticky tongue. [3 marks]

2. What is an extremophile?

   Give one example of how these organisms are well adapted to their habitats. [2 marks]

Studying ecosystems

Taking measurements in ecosystems

Ecosystems involve the interaction between **non-living** (**abiotic**) and **living** (**biotic**) parts of the environment. So it is important to identify which factors need to be measured in a particular habitat.

Abiotic factors include:
- light intensity
- temperature
- moisture levels
- soil pH and mineral content
- wind intensity and direction
- carbon dioxide levels for plants
- oxygen levels for aquatic animals.

Biotic factors include:
- availability of food
- new predators arriving
- new pathogens
- one species out-competing another.

Measuring biotic factors – sampling methods

It is usually impossible to count all the species living in a particular area, so a sample is taken.

When sampling, make sure you:
- **take a big enough sample** to make the estimate good and reliable – the larger the sample, the more accurate the results.
- **sample randomly** – the more random the sample, the more likely it is to be representative of the population.

Quadrats

Quadrats are square frames that typically have sides of length 0.5 m. They provide excellent results as long as they are placed randomly. The population of a certain species can then be estimated.

For example, if an average of 4 dandelion plants are found in a 0.25 m² quadrat, a scientist would estimate that 16 dandelion plants would be found in each 1 m² and 16 000 dandelion plants in a 1000 m² field.

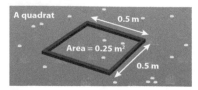

A quadrat · 0.5 m · Area = 0.25 m² · 0.5 m

Testing soil pH

Transects

Sometimes an environmental scientist may want to look at how species change across a habitat, or the boundary between two different habitats – for example, the plants found in a field as you move away from a hedgerow.

This needs a different approach that is systematic rather than random.

1. Lay down a line such as a tape measure. Mark regular intervals on it.
2. Next to the line, lay down a small quadrat. Estimate or count the number of plants of the different species. This can sometimes be done by estimating the percentage cover.
3. Move the quadrat along at regular intervals. Estimate and record the plant populations at each point until the end of the line.

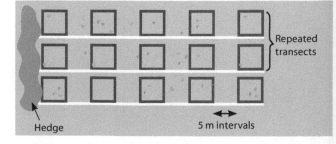

Hedge · Repeated transects · 5 m intervals

Sampling methods – animal populations

Sampling animal populations is more problematic as they are mobile and well adapted to evade capture. Here are four of the main techniques used.

Pooters

Insects sucked in here

You suck here

Fine mesh to stop you from sucking the insects into your mouth

This is a simple technique in which insects are gathered up easily without harm. With this method, you get to find out which species are actually present, although you have to be systematic about your sampling in order to get representative results and it is difficult to get ideas of numbers.

Sweepnets

Sweepnets are used in long grass or moderately dense woodland where there are lots of shrubs. Again, it is difficult to get truly representative samples, particularly in terms of the relative numbers of organisms.

Pitfall traps

Pitfall traps are set into the ground and used to catch small insects, e.g. beetles. Sometimes a mixture of ethanol or detergent and water is placed in the bottom of the trap to kill the samples, and prevent them from escaping. This method can give an indication of the relative numbers of organisms in a given area if enough traps are used to give a representative sample.

SUMMARY

- Ecosystems involve interaction between abiotic and biotic parts of the environment.
- Biotic factors can be measured by sampling methods such as quadrats and transects.

QUESTIONS

QUICK TEST

1. What information does a transect give you?

2. What is the difference between biotic and abiotic factors?

EXAM PRACTICE

1. A group of students were carrying out a quadrat survey to determine the population of daisies in a park. The park measured 60m x 90m.

   Their results are shown in the table.

Quadrat	1	2	3	4	5	6	7	8	9	10
No. daisies	5	2	1	0	4	5	2	0	6	3

   a) They placed the quadrats randomly.

      Why was this? **[1 mark]**

   b) Calculate the median number of daisies per quadrat. **[3 marks]**

   c) Another student calculated the mean number of daisies per quadrat to be 2.3.

      If the quadrats measured $1m^2 \times 1m^2$, calculate the estimated number of daisies in the park.

      Show your working. **[2 marks]**

Feeding relationships

Predator–prey relationships

Animals that kill and eat other animals are called **predators** (e.g. foxes, lynx). The animals that are eaten are called **prey** (e.g. rabbits, snowshoe hares).

Many animals can be both predator and prey. For instance, a stoat is a predator when it hunts rabbits and it is the prey when it is hunted by a fox.

Predator – stoat

Predator – fox

Prey – rabbit

Prey – stoat

In nature there is a delicate balance between the population of a predator (e.g. lynx) and its prey (e.g. snowshoe hare). However, the prey will always outnumber the predators.

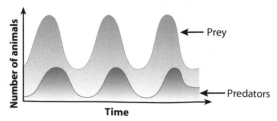

The number of predators and prey follow a classic population cycle. There will always be more hares than lynx and the population peak for the lynx will always come after the population peak for the hare. As the population cycle is cause and effect, they will always be out of phase.

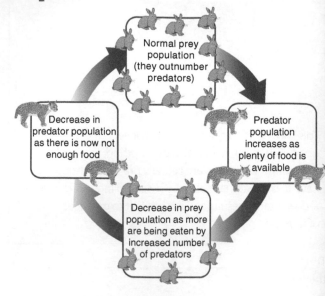

Normal prey population (they outnumber predators)

Predator population increases as plenty of food is available

Decrease in prey population as more are being eaten by increased number of predators

Decrease in predator population as there is now not enough food

Trophic levels

Communities of organisms are organised in an ecosystem according to their feeding habits.

Food chains show:

- the organisms that consume other organisms
- the transfer of **energy** and **materials** from organism to organism.

Energy from the Sun enters most food chains when green plants absorb sunlight to **photosynthesise**. Photosynthetic and chemosynthetic organisms are the producers of **biomass** for the Earth. Feeding passes this energy and biomass from one organism to the next along the food chain.

A food chain

Green plant:
producer

Rabbit:
primary consumer

Stoat:
secondary consumer

Fox:
tertiary consumer

The arrow shows the flow of energy and biomass along the food chain.

● All food chains start with a **producer**.

● The rabbit is a herbivore (plant eater), also known as the **primary consumer**.

● The stoat is a carnivore (meat eater), also known as the **secondary consumer**.

● The fox is the top carnivore in this food chain, the **tertiary consumer**.

Each consumer or producer occupies a **trophic level** (feeding level).

● Level 1 are producers.

● Level 2 are primary consumers.

● Level 3 are secondary consumers.

● Level 4 are tertiary consumers.

Excretory products and uneaten parts of organisms can be the starting points for other food chains, especially those involving decomposers.

Decomposers

SUMMARY

● Predators kill and eat prey. The prey will always outnumber predators.

● Food chains show the organisms that consume other organisms, and how energy and materials are transferred from organism to organism.

● A trophic level is a feeding level in a food chain.

QUESTIONS

QUICK TEST

1. What is a predator?

2. What is meant by a trophic level?

3. How does energy from the sun enter most food chains?

4. In the food chain shown above, what is the producer?

EXAM PRACTICE

1. The food web shown below exists in a freshwater pond habitat:

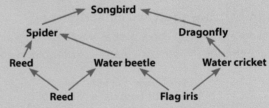

Write out two food chains, one involving three organisms and the other involving four organisms.

Each chain must include a producer, one primary consumer and one secondary consumer.

The second chain must also include an apex predator. **[2 marks]**

a) _____ ⟶ _____ ⟶ _____

b) _____ ⟶ _____ ⟶ _____ ⟶ _____

2. Rabbits and stoat numbers tend to follow a cyclical pattern as their numbers change over the years.

Explain why this is so. **[3 marks]**

Environmental change & biodiversity

Environmental change

Waste management

The human population is increasing exponentially (i.e. at a rapidly increasing rate). This is because birth rates exceed death rates by a large margin.

So the use of finite resources like fossil fuels and minerals is accelerating. In addition, waste production is going up:

- **on land**, from domestic waste in landfill, toxic chemical waste, pesticides and herbicides
- **in water**, from sewage fertiliser and toxic chemicals
- **in the air**, from smoke, carbon dioxide and sulfur dioxide.

Acid rain

When coal or oils are burned, sulfur dioxide is produced. Sulfur dioxide and nitrogen dioxide dissolve in water to produce acid rain.

Acid rain can:

- damage trees, stonework and metals
- make rivers and lakes acidic, which means some organisms can no longer survive.

The acids can be carried a long way away from the factories where they are produced. Acid rain falling in one country could be the result of fossil fuels being burned in another country.

The greenhouse effect and global warming

The diagram explains how global warming can lead to climate change. This in turn leads to lower biodiversity.

Small amount of infra-red radiation transmitted into space

Infra-red radiation re-radiated into atmosphere

Carbon dioxide and methane trap the infra-red radiation (heat) as the longer wavelengths are not transmitted as easily.

UV rays from the sun reach Earth and are absorbed

The consequences of global warming are:

- a rise in sea levels leading to flooding in low-lying areas and loss of habitat
- the migration of species and changes in their distribution due to more extreme temperature and rainfall patterns; some organisms won't survive being displaced into new habitats, or newly migrated species may outcompete native species. The overall effect is a loss of biodiversity.

WS You may be asked to evaluate methods used to address problems caused by human impact on the environment.

For example, here are some figures relating to quotas and numbers of haddock in the North Sea in two successive years.

	2009	2010
Haddock quota (tonnes)	27 507	23 381
Estimated population (thousands)	102	101

What conclusions could you draw from this data? What additional information would you need to give a more accurate picture?

Biodiversity

Biodiversity is a measure of the number and variety of species within an ecosystem. A healthy ecosystem:
- has a large biodiversity
- has a large degree of interdependence between species
- is stable.

Species depend on each other for food, shelter and keeping the external physical environment maintained. Humans have had a negative impact on biodiversity due to:
- pollution killing plants and animals
- degrading the environment through deforestation and removing resources such as minerals and fossil fuels
- over-exploiting habitats and organisms.

Only recently have humans made efforts to reduce their impact on the environment. It is recognised that maintaining biodiversity is important to ensure the continued survival of the human race.

Impact of land use

As humans increase their economic activity, they use land that would otherwise be inhabited by living organisms. Examples of habitat destruction include:
- farming
- quarrying
- dumping waste in landfill.

A marble quarry

Peat bogs

Peat bogs are important habitats. They support a wide variety of organisms and act as carbon sinks.

If peat is burned it releases carbon dioxide into the atmosphere and contributes to global warming. Removing peat for use as compost in gardens takes away the habitat for specialised animals and plants that aren't found in other habitats.

Deforestation

Deforestation is a particular problem in tropical regions. Tropical rainforests are removed to:

- **release land for cattle and rice fields** – these are needed to feed the world's growing population and for increasingly Western-style diets

- **grow crops for biofuel** – the crops are converted to **ethanol-based** fuels for use in petrol and diesel engines. Some specialised engines can run off pure ethanol.

The consequences of deforestation are:

- There are fewer plants, particularly trees, to absorb carbon dioxide. This leads to increased carbon dioxide in the atmosphere and accelerated global warming.
- Combustion and decay of the wood from deforestation releases more carbon dioxide into the atmosphere.
- There is reduced biodiversity as animals lose their habitats and food sources.

Maintaining biodiversity

To prevent further losses in biodiversity and to improve the balance of ecosystems, scientists, the government and environmental organisations can take action.

- Scientists establish **breeding programmes** for **endangered species**. These may be captive methods where animals are enclosed, or protection schemes that allow rare species to breed without being poached or killed illegally.

- The government sets **limits** on **deforestation** and **greenhouse gas emissions**.

Environmental organisations:

- **protect and regenerate** shrinking habitats such as mangrove swamps, heathlands and coral reefs

- **conserve and replant** hedgerows around the margins of fields used for crop growth

- introduce and encourage **recycling** initiatives that reduce the volume of landfill.

SUMMARY

- **The human population is increasing exponentially. Waste production is increasing.**
- **Burning of fossil fuels contributes to acid rain and global warming.**
- **Biodiversity is the number and variety of species in an ecosystem. Maintaining biodiversity is vital.**

QUESTIONS

QUICK TEST

1. What is meant by deforestation?

2. Name one pollutant gas that contributes to acid rain.

3. What type of habitat acts as a carbon sink?

EXAM PRACTICE

1. In Ireland, four species of bumble bee are now endangered. Scientists are worried that numbers may become so low that they are inadequate to provide pollination to certain plants.

   a) Explain how the disappearance of these four species of bumble bee might affect biodiversity as a whole. **[3 marks]**

   b) There are many threats to bumble bees, such as the possible effect of pesticides such as *neonicotinoids*; loss of suitable habitat, disappearance of wildflowers and new diseases.

   Suggest two conservation measures that might help increase numbers of bumble bees in the wild. **[2 marks]**

Recycling and sustainability

The water cycle

Water is a vital part of the biosphere. Most organisms consist of over 50% water.

The two key processes that drive the water cycle are **evaporation** and **condensation**.

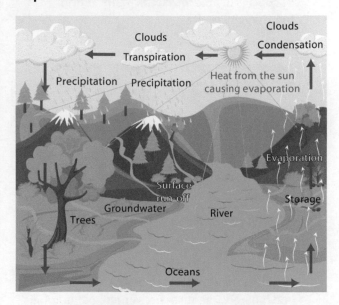

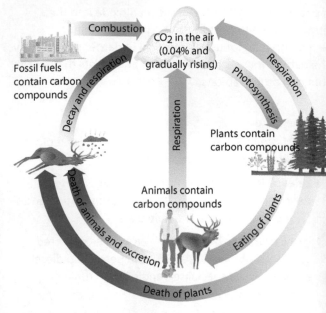

The carbon cycle

The constant recycling of carbon is called the carbon cycle.

- Carbon dioxide is removed from the atmosphere by green plants for photosynthesis.

- Plants and animals respire, releasing carbon dioxide into the atmosphere.

- Animals eat plants and other animals, which incorporates carbon into their bodies. In this way, carbon is passed along food chains and webs.

- Microorganisms such as fungi and bacteria feed on dead plants and animals, causing them to decay. The microorganisms respire and release carbon dioxide gas into the air. Mineral ions are returned to the soil through decay. This extraction and return of nutrients to the soil is called the decay cycle.

- Some organisms' bodies are turned into fossil fuels over millions of years, trapping the carbon as coal, peat, oil and gas.

- When fossil fuels are burned (combustion), the carbon dioxide is returned to the atmosphere.

Eutrophication

Overusing fertilisers in intensive farming can lead to eutrophication.

1 Fertilisers or sewage can run into the water and pollute it. As a result, there are a lot of nitrates and phosphates, which leads to rapid growth of algae.

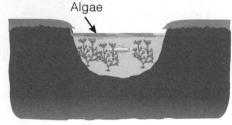

2 The algal blooms reproduce quickly, then die and rot. They also block off sunlight, which causes underwater plants to die and rot.

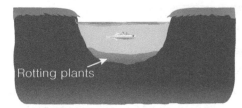

3 The number of aerobic bacteria increase. As they feed on the dead organisms they use up oxygen. This causes larger organisms and plants to die because they are unable to respire.

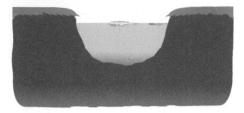

Sustainable fishing

Fish stocks are declining across the globe. The problem is so big that unless methods are used to halt the decline, some species (such as the northwest Atlantic cod) might disappear.

Methods to make fishing sustainable include:

● governments imposing **quotas** that limit the weight of fish that can be taken from the oceans on a yearly basis – this practice is not always followed and illegal fishing is difficult to combat

● increasing the mesh size of nets to allow smaller fish to escape and reach adulthood, so that they can breed.

SUMMARY

● **Materials within ecosystems are constantly being recycled and used to provide the substances that make up future organisms.**

● **Two of these substances are water and carbon.**

● **Intensive farming produces high yields but also has drawbacks, including questions around animal welfare, and the overuse of fertilisers causing eutrophication.**

QUESTIONS

QUICK TEST

1. Which two processes drive the water cycle?

2. How much water do most organisms consist of?

3. Explain why governments impose quotas on fishing.

EXAM PRACTICE

1. Carbon is recycled in the environment in a process called the carbon cycle. The main processes of the carbon cycle are shown below:

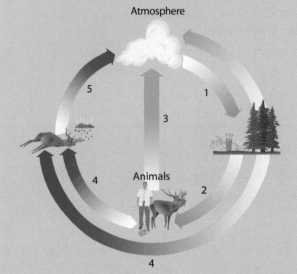

a) Name the process that occurs at stage 1 in the diagram. **[1 mark]**

b) i) Name the process that occurs at Stage 4 in the diagram. **[1 mark]**

ii) Describe how this process leads to the release of carbon dioxide into the atmosphere. **[2 marks]**

2. Environmental groups have criticised governments of different countries that they have not done enough to halt the decline of fish stocks in the North Sea.

Explain the advantages and disadvantages of applying quotas and increased mesh sizes as a policy. **[2 marks]**

Atoms and the periodic table

Atoms and elements

- All substances are made of **atoms**.
- An atom is the smallest part of an **element** that can exist.
- There are approximately 100 different elements, all of which are shown in the **periodic table**.

Below is part of the periodic table, showing the names and symbols of the first 20 elements.

Atoms of each element are represented by a chemical symbol in the periodic table.

Some elements have more than one letter in their symbol. The first letter is always a capital and the other letter is lower case.

Chemical equations

We can use word or symbol equations to represent chemical reactions.

Word equation:

> sodium + oxygen ⟶ sodium oxide

Symbol equation:

> $4Na + O_2 \longrightarrow 2Na_2O$

1 hydrogen **H**								2 helium **He**
3 lithium **Li**	4 beryllium **Be**	5 boron **B**	6 carbon **C**	7 nitrogen **N**	8 oxygen **O**	9 fluorine **F**	10 neon **Ne**
11 sodium **Na**	12 magnesium **Mg**	13 aluminium **Al**	14 silicon **Si**	15 phosphorus **P**	16 sulfur **S**	17 chlorine **Cl**	18 argon **Ar**
19 potassium **K**	20 calcium **Ca**						

The symbol for magnesium is **Mg**

The symbol for the element oxygen is **O**

Elements, compounds and mixtures

Element		Elements contain only one type of atom and consist of single atoms or atoms bonded together
Compound		Compounds contain two (or more) elements that are chemically combined in fixed proportions
Mixture		Mixtures contain two or more elements (or compounds) that are together but not chemically combined

Separating mixtures

Compounds can only be separated by chemical reactions but **mixtures** can be separated by physical processes (not involving chemical reactions) such as filtration, crystallisation, simple (and fractional) distillation and chromatography.

How different mixtures can be separated	
Method of separation	**Diagram**
Filtration Used for separating insoluble solids from liquids, e.g. sand from water	Filter paper / Filter funnel / Sand and water / Sand / Beaker / Clear water (filtrate)
Crystallisation Used to separate solids from solutions, e.g. obtaining salt from salty water	Evaporating dish / Mixture / Wire gauze / Tripod stand / Bunsen burner
Simple distillation Simple distillation is used to separate a liquid from a solution, e.g. water from salty water	Thermometer / Round-bottomed flask / Water out / Liebig condenser / Water in / Heat
Fractional distillation Fractional distillation is used to separate liquids that have different boiling points, e.g. ethanol and water	
Chromatography Used to separate dyes, e.g. the different components of ink	Solvent front / Separated dyes / Chromatography paper / Ink spots / Pencil line / Solvent

SUMMARY

● **An atom is the smallest part of an element.**
● **The elements are shown in the periodic table.**
● **Mixtures can be separated by physical processes; compounds must be separated by chemical reactions.**

QUESTIONS

QUICK TEST

1. What are all substances made from?

2. What name is given to a substance that contains atoms of different elements chemically joined together?

3. Which method of separation would you use to obtain salt from salt water?

EXAM PRACTICE

1. A student is provided with a mixture of magnesium chloride and magnesium oxide. Magnesium chloride is soluble in water but magnesium oxide is insoluble.

   a) Write a balanced symbol equation for the reaction between magnesium and oxygen to form magnesium oxide.

   Include state symbols in your answer. **[2 marks]**

   b) What is meant by the term 'mixture'? **[1 mark]**

   c) Describe how the mixture of magnesium chloride and magnesium oxide could be separated. **[3 marks]**

Atomic structure

(ws) Scientific models of the atom

Scientists had originally thought that atoms were tiny spheres that could not be divided.

John Dalton conducted experiments in the early 19th century and concluded that...
- all matter is made of indestructible atoms
- atoms of a particular element are identical
- atoms are rearranged during chemical reactions
- compounds are formed when two or more different types of atom join together.

Upon discovery of the electron by **J. J. Thomson** in 1897, the 'plum pudding' model suggested that the atom was a ball of positive charge with negative electrons embedded throughout.

The results from **Rutherford**, **Geiger** and **Marsden's** alpha scattering experiments (1911–1913) led to the plum pudding model being replaced by the nuclear model.

In this experiment, alpha particles (which are positive) are fired at a thin piece of gold. A few of the alpha particles do not pass through the gold and are deflected. Most went straight through the thin piece of gold. This led Rutherford, Geiger and Marsden to suggest that this is because the positive charge of the atom is confined in a small volume (now called the nucleus).

Niels Bohr adapted the nuclear model in 1913, by suggesting that electrons orbit the nucleus at specific distances. Bohr's theoretical calculations were backed up by experimental results.

Later experiments led to the idea that the positive charge of the nucleus was subdivided into smaller particles (now called protons), with each particle having the same amount of positive charge.

The work of **James Chadwick** suggested in 1932 that the nucleus also contained neutral particles that we now call neutrons.

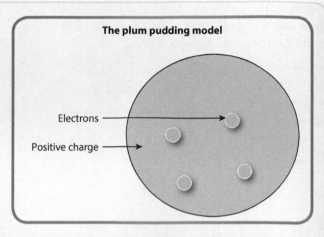

The plum pudding model

Electrons

Positive charge

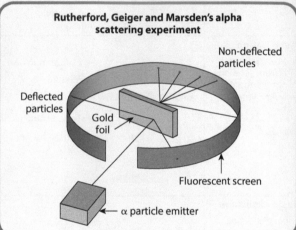

Rutherford, Geiger and Marsden's alpha scattering experiment

Non-deflected particles

Deflected particles

Gold foil

Fluorescent screen

α particle emitter

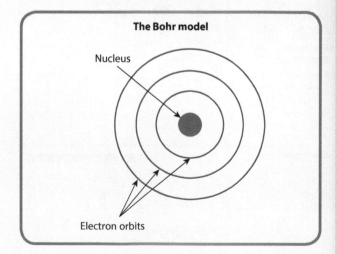

The Bohr model

Nucleus

Electron orbits

Properties of atoms

Particle	Relative charge	Relative mass
Proton	+1	1
Neutron	0	1
Electron	−1	negligible

Atoms have a neutral charge. This is because the number of protons is equal to the number of electrons.
- Atoms of different elements have different numbers of protons. This number is called the atomic number.
- Atoms are very small, having a radius of approximately 0.1 nm (1×10^{-10} m).
- The radius of the nucleus is approximately $\frac{1}{10\,000}$ of the size of the atom.

> The **mass number** tells you the total number of protons and neutrons in an atom

$$^{23}_{11}\text{Na}$$

> The **atomic number** tells you the number of protons in an atom

> mass number − atomic number = number of neutrons

Some atoms can have different numbers of neutrons. These atoms are called **isotopes**. The existence of isotopes results in the relative atomic mass of some elements, e.g. chlorine, not being whole numbers.

> **HT** Chlorine exists as two isotopes. Chlorine-35 makes up 75% of all chlorine atoms. Chlorine-37 makes up the other 25%. We say that the abundance of chlorine-35 is 75%.
>
> The relative atomic mass of chlorine can be calculated as follows:
>
> $$\frac{(\text{mass of isotope 1} \times \text{abundance}) + (\text{mass of isotope 2} \times \text{abundance})}{100}$$
>
> $$= \frac{(35 \times 75) + (37 \times 25)}{100} = 35.5$$

SUMMARY

- Notable scientists who worked on atomic structure are John Dalton; J. J. Thompson; Rutherford, Geiger and Marsden; Niels Bohr and James Chadwick.
- Atoms have protons (positive), neutrons (neutral) and electrons (negative) and are neutral.

QUESTIONS

QUICK TEST

1. What is the relative charge of a proton?
2. How many protons, neutrons and electrons are present in the following atom?

$$^{24}_{12}\text{Mg}$$

3. What name is given to atoms of the same element which have the same number of protons but different numbers of neutrons?

EXAM PRACTICE

1. The diagram below shows J. J. Thomson's plum pudding model of an atom and Bohr's nuclear model of an atom.

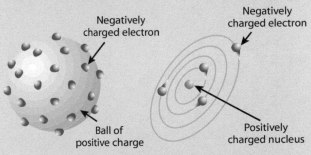

Describe the key differences between these two different models of the atom. **[3 marks]**

Electronic structure & the periodic table

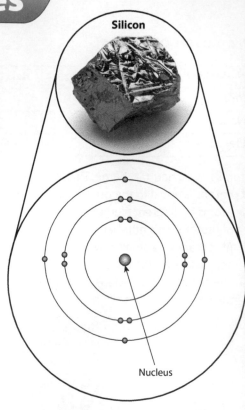

Silicon

Nucleus

Electronic structures

Electrons in an atom occupy the lowest available energy level (shell).

The first energy level (closest to the nucleus) can hold up to two electrons. The second and third energy levels can hold up to eight electrons.

For example, silicon has the atomic number 14. This means that there are 14 protons in the nucleus of a silicon atom and therefore there must be 14 electrons (so that the atom is neutral).

The electronic structure of silicon can be written as 2, 8, 4 or shown in a diagram like the one on the right.

Silicon is in group 4 of the periodic table. This is because it has four electrons in its outer shell. The chemical properties (reactions) of an element are related to the number of electrons in the outer shell of the atom.

The electronic structure of the first 20 elements is shown here.

Group 1	Group 2			Hydrogen, H Atomic No. = 1 No. of electrons = 1	Group 3	Group 4	Group 5	Group 6	Group 7	Group 8
										Helium, He Atomic No. = 2 No. of electrons = 2
				1						2
Lithium, Li Atomic No. = 3 No. of electrons = 3	Beryllium, Be Atomic No. = 4 No. of electrons = 4				Boron, B Atomic No. = 5 No. of electrons = 5	Carbon, C Atomic No. = 6 No. of electrons = 6	Nitrogen, N Atomic No. = 7 No. of electrons = 7	Oxygen, O Atomic No. = 8 No. of electrons = 8	Fluorine, F Atomic No. = 9 No. of electrons = 9	Neon, Ne Atomic No. = 10 No. of electrons = 10
2, 1	2, 2				2, 3	2, 4	2, 5	2, 6	2, 7	2, 8
Sodium, Na Atomic No. = 11 No. of electrons = 11	Magnesium, Mg Atomic No. = 12 No. of electrons = 12				Aluminium, Al Atomic No. = 13 No. of electrons = 13	Silicon, Si Atomic No. = 14 No. of electrons = 14	Phosphorus, P Atomic No. = 15 No. of electrons = 15	Sulfur, S Atomic No. = 16 No. of electrons = 16	Chlorine, Cl Atomic No. = 17 No. of electrons = 17	Argon, Ar Atomic No. = 18 No. of electrons = 18
2, 8, 1	2, 8, 2				2, 8, 3	2, 8, 4	2, 8, 5	2, 8, 6	2, 8, 7	2, 8, 8
Potassium, K Atomic No. = 19 No. of electrons = 19	Calcium, Ca Atomic No. = 20 No. of electrons = 20									
2, 8, 8, 1	2, 8, 8, 2									

THE TRANSITION METALS

This table is arranged in order of atomic (proton) numbers, placing the elements in groups. Elements in the same group have the same number of electrons in their highest occupied energy level (outer shell).

The electron configuration of oxygen is 2, 6 because there are...
- 2 electrons in the first shell
- 6 electrons in the second shell.

The periodic table

The elements in the periodic table are arranged in order of increasing atomic (proton) number.

The table is called a **periodic table** because similar properties occur at regular intervals.

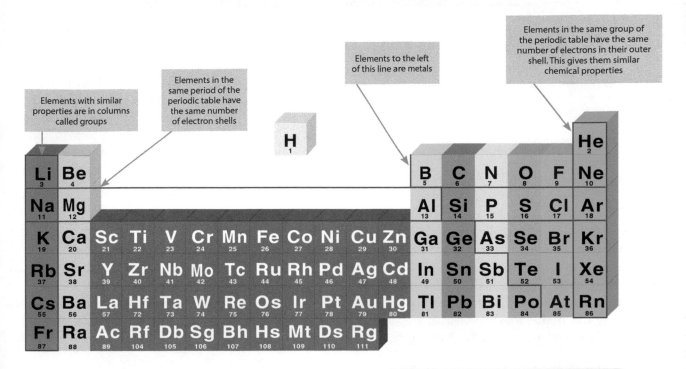

Elements with similar properties are in columns called groups

Elements in the same period of the periodic table have the same number of electron shells

Elements to the left of this line are metals

Elements in the same group of the periodic table have the same number of electrons in their outer shell. This gives them similar chemical properties

Development of the periodic table

Before the discovery of protons, neutrons and electrons, early attempts to classify the elements involved placing them in order of their atomic weights. These early attempts resulted in incomplete tables and the placing of some elements in appropriate groups based on their chemical properties.

Dmitri Mendeleev overcame some of these problems by leaving gaps for elements that he predicted were yet to be discovered. He also changed the order for some elements based on atomic weights. Knowledge of isotopes made it possible to explain why the order based on atomic weights was not always correct.

Metals and non-metals

● Metals are elements that react to form positive ions.

● Elements that do not form positive ions are non-metals.

Typical properties of metals and non-metals	
Metals	**Non-metals**
Have high melting / boiling points	Have low melting / boiling points
Conduct heat and electricity	Thermal and electrical insulators
React with oxygen to form alkalis	React with oxygen to form acids
Shiny	Dull
Malleable and ductile	Brittle

SUMMARY

● **Elements in the periodic table are arranged in order of increasing atomic number.**

● **Dmitri Mendeleev left gaps in the periodic table for elements yet to be discovered.**

QUESTIONS

QUICK TEST

1. Potassium has the atomic number 19. What is the electronic structure of potassium?

2. Why did Mendeleev leave gaps in his periodic table?

EXAM PRACTICE

1. Element X is in period 3 and group 2 of the periodic table. Element Y has 14 protons in each atom.

   a) Give the electronic configuration of element X. **[1 mark]**

   b) Will element Y be before or after element X in the periodic table?

   Explain your answer. **[2 marks]**

Groups 0, 1 and 7

Group 0

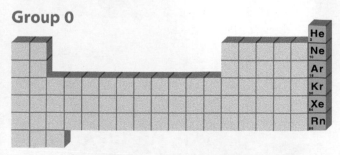

The elements in group 0 are called the **noble gases**. They are chemically inert (unreactive) and do not easily form molecules because their atoms have full outer shells (energy levels) of electrons. The inertness of the noble gases, combined with their low density and non-flammability, means that they can be used in airships, balloons, light bulbs, lasers and advertising signs.

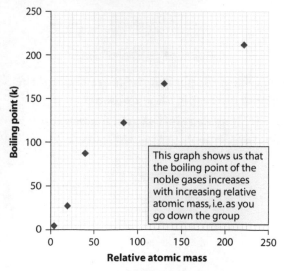

This graph shows us that the boiling point of the noble gases increases with increasing relative atomic mass, i.e. as you go down the group

Group 1 (the alkali metals)

The alkali metals…

- have a low density (lithium, sodium and potassium float on water)
- react with non-metals to form ionic compounds in which the metal ion has a charge of +1
- form compounds that are white solids and dissolve in water to form colourless solutions.

When added to water, the alkali metals float, move about on the surface of the water and effervesce. Sodium melts and potassium reacts with a lilac flame being observed. Hydrogen gas and metal hydroxides are formed. The metal hydroxides dissove in water to form alkaline solutions. For example, for the reaction between sodium and water:

$$2Na_{(s)} + 2H_2O_{(l)} \rightarrow 2NaOH_{(aq)} + H_{2(g)}$$

The alkali metals become more reactive as you go down the group because the outer shell gets further away from the positive attraction of the nucleus. This makes it easier for an atom to lose an electron from its outer shell.

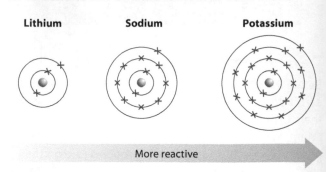

More reactive

Group 7 (the halogens)

The halogens…

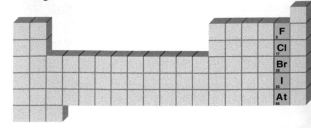

- are non-metals
- consist of diatomic molecules (molecules made up of two atoms)
- react with metals to form ionic compounds where the halide ion has a charge of −1
- form molecular compounds with other non-metals
- form hydrogen halides (e.g. HCl), which dissolve in water, forming acidic solutions.

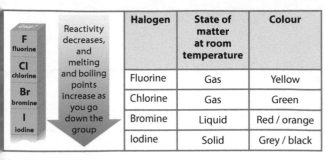

Halogen	State of matter at room temperature	Colour
Fluorine	Gas	Yellow
Chlorine	Gas	Green
Bromine	Liquid	Red / orange
Iodine	Solid	Grey / black

Halogens become less reactive as you go down the group because the outer electron shell gets further away from the attraction of the nucleus, and so an electron is gained less easily.

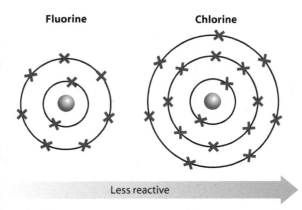

Fluorine   Chlorine

Less reactive

Displacement reactions of halogens

A more reactive halogen will displace a less reactive halogen from an aqueous solution of its metal halide.

For example:

chlorine + potassium bromide → potassium chloride + bromine

Cl_2 +   2KBr   →   2KCl   + Br_2

The products of reactions between halogens and aqueous solutions of halide ion salts are as follows.

	Halide salts		
	Potassium chloride, KCl	Potassium bromide, KBr	Potassium iodide, KI
Chlorine, Cl_2	No reaction	Potassium chloride + bromine	Potassium chloride + iodine
Bromine, Br_2	No reaction	No reaction	Potassium bromide + iodine
Iodine, I_2	No reaction	No reaction	No reaction

(left side label: Halogens)

SUMMARY

- The elements in group 0 are the noble gases; they are chemically inert.
- Group 1 elements are the alkali metals; they get more reactive as you go down the group.
- Group 7 elements are the halogens; they are non-metals which get less reactive as you go down the group.

QUESTIONS

QUICK TEST

1. Suggest two uses for the noble gases.

2. What are the products of the reaction between potassium and water?

3. What is the trend in reactivity of the halogens as you go down the group?

EXAM PRACTICE

1. A small piece of sodium is added to a trough of water.

   a) What will be observed during this reaction?
   [3 marks]

   b) Write a balanced symbol equation for the reaction occurring.

   Include state symbols.
   [2 marks]

2. Chlorine gas is bubbled through sodium bromide solution.

   a) Explain why the colourless solution turns brown after the chlorine gas has been bubbled through. [2 marks]

   b) Explain why this displacement reaction occurs. [2 marks]

Chemical bonding

There are three types of chemical bond:

● ionic
● covalent
● metallic

Ionic bonding

Ionic bonds occur between metals and non-metals. An ionic bond is the electrostatic force of attraction between two oppositely charged ions (called cations and anions).

Ionic bonds are formed when metal atoms transfer electrons to non-metal atoms. This is done so that each atom forms an ion with a full outer shell of electrons.

Example 1

The formation of an ionic bond between sodium and chlorine

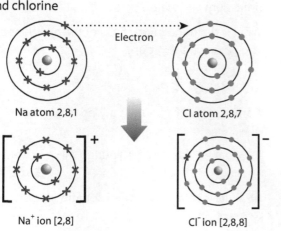

Na atom 2,8,1

Cl atom 2,8,7

Electron

Na⁺ ion [2,8]

Cl⁻ ion [2,8,8]

Example 2

The formation of the ionic bond between magnesium and oxygen

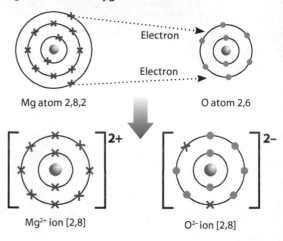

Electron

Electron

Mg atom 2,8,2

O atom 2,6

Mg²⁺ ion [2,8]

O²⁻ ion [2,8]

Covalent bonding

Covalent bonds occur between two non-metal atoms. Atoms share electrons so that each atom ends up with a full outer shell of electrons. A covalent bond is a shared pair of electrons, e.g.:

● the formation of a covalent bond between two hydrogen atoms

Hydrogen atoms

A hydrogen molecule

Covalent bond

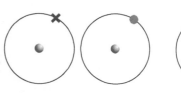

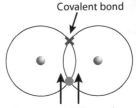

Outermost shells overlap

● the covalent bonding in methane, CH_4.

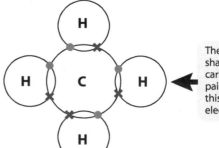

The hydrogen atoms share electrons. The carbon atom shares four pairs of electrons. They do this by sharing a pair of electrons in a single bond

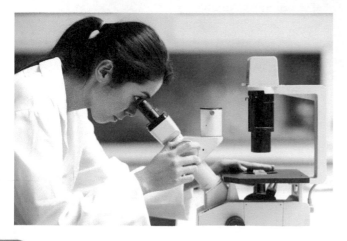

Double covalent bonds occur when two pairs of electrons are shared between atoms, for example in carbon dioxide.

Carbon dioxide

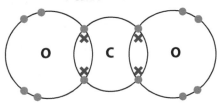

Metallic bonding

Metals consist of giant structures. Each atom loses its outer shell electrons and these electrons become delocalised, i.e. they are free to move through the structure.

The metal cations are arranged in a regular pattern called a lattice.

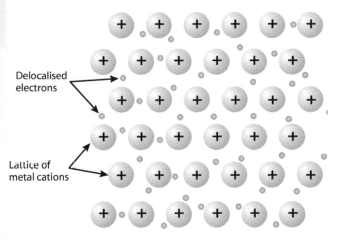

Delocalised electrons

Lattice of metal cations

QUESTIONS

QUICK TEST

1. What type of bonding occurs between non-metal atoms?

2. How many electrons are present in every covalent bond?

3. What is a delocalised electron?

4. What is an ionic bond?

EXAM PRACTICE

1. The electronic configurations of atoms of potassium and sulfur are:

   K 2,8,8,1

   S 2,8,6

   a) Describe the changes in the electronic configurations of potassium and sulfur when these atoms react to form potassium sulfide. **[3 marks]**

   b) Give the formula of potassium sulfide. **[1 mark]**

2. Ammonia (NH_3) contains covalent bonds.

   a) Draw a dot and cross diagram to show the covalent bonding in ammonia. Show the outer electrons only. **[2 marks]**

   b) Explain why neon does not form covalent or ionic bonds. **[2 marks]**

3. Thallium is a metal in group 3 of the periodic table.

   Describe the structure of thallium. **[2 marks]**

SUMMARY

- There are three types of chemical bond: ionic, covalent and metallic.
- Ionic bonds occur between metals and non-metals.
- Covalent bonds occur between non-metals.

Ionic and covalent structures

Structure of ionic compounds

Compounds containing ionic bonds form giant structures. These are held together by strong electrostatic forces of attraction between the oppositely charged ions. These forces act in all directions throughout the lattice.

The diagram below represents a typical giant ionic structure, sodium chloride.

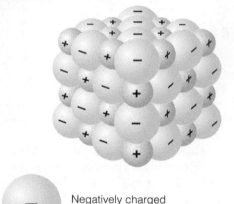

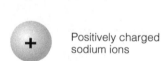

Negatively charged chloride ions

Positively charged sodium ions

The ratio of each ion present in the structure allows the empirical formula of the compound to be worked out. In the diagram above, there are equal numbers of sodium ions and chloride ions. This means that the empirical formula is NaCl.

Sodium chloride

Structure of covalent compounds

Covalently bonded substances may consist of....

● small molecules / simple molecular structures (e.g. Cl_2, H_2O and CH_4)

● large molecules (e.g. polymers)

● giant covalent structures (e.g. diamond, graphite and silicon dioxide).

Simple molecular structures

The bonding between hydrogen and carbon in methane can be represented in several ways, as shown here.

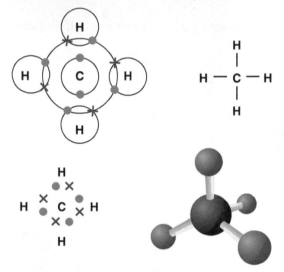

Polymers

Polymers can be represented in the form:

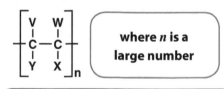

where *n* is a large number

V, W, X and Y represent the atoms bonded to the carbon atoms

For example, poly(ethene) can be represented as:

Giant covalent structures

This is the giant covalent structure of silicon dioxide.

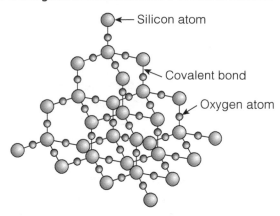

The formula of silicon dioxide is SiO_2 – this can be deduced by looking at the ratio of Si to O atoms in the diagram above.

Models, such as dot-and-cross diagrams, ball-and-stick diagrams and two- / three-dimensional diagrams to represent structures, are limited in value, as they do not accurately represent the structures of materials. For example, a chemical bond is not a solid object as depicted in some models, it is actually an attraction between particles. The relative size of different atoms is often not shown when drawing diagrams.

Diamond – a giant covalent structure

Graphite – a giant covalent structure

SUMMARY

- **Ionic compounds are held together by strong electrostatic forces of attraction between oppositely charged ions.**
- **Substances with covalent bonding can either have simple molecular, giant covalent or polymer structures.**

QUESTIONS

QUICK TEST

1. Name a compound that has a giant ionic structure.

2. What forces hold ionic structures together?

3. What are the three types of covalent substance?

EXAM PRACTICE

1. The diagram below represents one way that the structure of an ionic lattice can be modelled.

   Key

    metal X ion

    chloride (Cl^-) ion

   a) What is the empirical formula of X chloride?

      Explain how you worked this out. **[2 marks]**

   b) Give one way in which the model shown is limited in value. **[1 mark]**

2. Carbon dioxide and silicon dioxide are both oxides of group 4 elements. They contain the same type of chemical bond but they have different structures.

   a) Name the type of bonding present in both carbon dioxide and silicon dioxide. **[1 mark]**

   b) What type of structure does carbon dioxide have? **[1 mark]**

   c) What type of structure does silicon dioxide have? **[1 mark]**

States of matter: properties of compounds

States of matter

The three main states of matter are **solids**, **liquids** and **gases**. Individual atoms do not have the same properties as these bulk substances. The diagram below shows how they can be interconverted and also how the particles in the different states of matter are arranged.

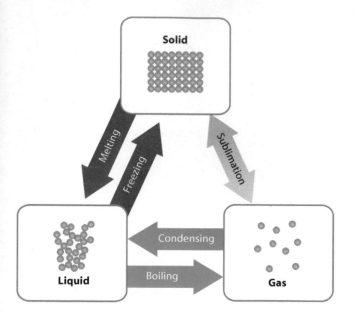

These are physical changes because the particles are either gaining or losing energy and are not undergoing a chemical reaction. Particles in a gas have more energy than in a liquid; particles in a liquid have more energy than in a solid.

Melting and freezing occur at the same temperature. Condensing and boiling also occur at the same temperature. The amount of energy needed to change state depends on the strength of the forces between the particles of the substance.

The stronger the forces between the particles, the higher the melting and boiling points of the substance.

Properties of ionic compounds	
Property	**Explanation**
High melting and boiling points	There are lots of strong bonds throughout an ionic lattice which require lots of energy to break.
Electrical conductivity	Ionic compounds conduct electricity when molten or dissolved in water because the ions are free to move and carry the charge. Ionic solids do not conduct electricity because the ions are in a fixed position and are unable to move.

HT The model on the left is limited in value because….

● it does not indicate that there are forces between the spheres

● all particles are represented as spheres

● the spheres are solid.

Properties of small molecules

Substances made up of small molecules are usually gases or liquids at room temperature. They have relatively low melting and boiling points because there are weak (intermolecular) forces that act between the molecules. It is these weak forces and not the strong covalent bonds that are broken when the substance melts or boils.

Substances made up of small molecules do not normally conduct electricity. This is because the molecules do not have an overall electric charge or delocalised electrons.

Polymers

Polymers are very large molecules made up of atoms joined together by strong covalent bonds. The intermolecular forces between polymer molecules are much stronger than in small molecules because the molecules are larger. This is why most polymers are solid at room temperature.

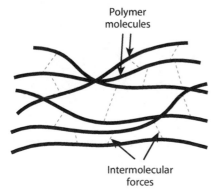

Polythene

Polymer molecules

Intermolecular forces

Giant covalent structures

Substances with a giant covalent structure are solids at room temperature. They have relatively high melting and boiling points. This is because there are lots of strong covalent bonds that need to be broken.

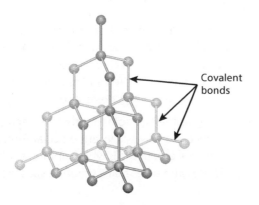

Covalent bonds

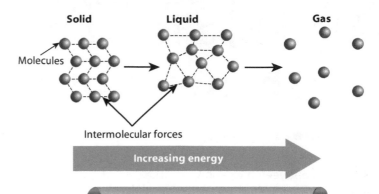

Solid   Liquid   Gas

Molecules

Intermolecular forces

Increasing energy

SUMMARY

● The three main states of matter are solids, liquids and gases.

● Polymers are large molecules made up of atoms joined by strong covalent bonds; most are solids at room temperature.

● Giant covalent structures are solids at room temperature.

QUESTIONS

QUICK TEST

1. What are the three states of matter?

2. What needs to be broken in order to melt a substance made up of small molecules such as water?

EXAM PRACTICE

1. Phosphorus tribromide (PBr_3) has a melting point of -40°C and a boiling point of 175°C.

   a) At what temperature does liquid phosphorus tribromide turn into a solid? **[1 mark]**

   b) What state of matter is phosphorus tribromide at 100°C? **[1 mark]**

   c) Name the physical process occurring when the temperature of phosphorus decreases from 200°C to 175°C. **[1 mark]**

2. Sodium chloride does not conduct electricity at room temperature but does when molten.

   Explain this observation. **[3 marks]**

Metals, alloys & the structure and bonding of carbon

Structure and properties of metals

Metals have giant structures. Metallic bonding (the attraction between the cations and the delocalised electrons) is strong, meaning that most metals have high melting and boiling points.

The layers are able to slide over each other, which means that metals can be bent and shaped.

Metals are good conductors of electricity because the delocalised electrons are able to move.

The delocalised electrons also transfer energy meaning that they are good thermal conductors.

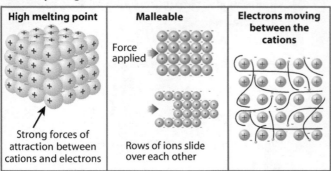

High melting point	Malleable	Electrons moving between the cations
Strong forces of attraction between cations and electrons	Force applied — Rows of ions slide over each other	

Alloys

Most metals we use are alloys. Many pure metals (such as gold, iron and aluminium) are too soft for many uses and so are mixed with other materials (usually metals) to make alloys.

The different sizes of atoms in alloys make it difficult for the layers to slide over each other. This is why alloys are harder than pure metals.

Typical alloy structure

Metal   Other atoms

Structure and bonding of carbon

Carbon has four different structures:

- diamond
- graphite
- graphene
- fullerenes

Diamond

Diamond has a giant covalent (macromolecular) structure where each carbon atom is bonded to four others.

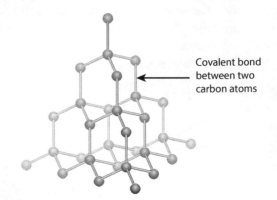

Covalent bond between two carbon atoms

In diamond, there are lots of very strong covalent bonds so diamond…

- is hard
- has a high melting point.

For these reasons, diamond is used in making cutting tools.

There are no free electrons in diamond so it does not conduct electricity.

Graphite

Graphite also has a giant covalent structure, with each carbon atom forming three covalent bonds, resulting in layers of hexagonal rings of carbon atoms. Carbon has four electrons in its outer shell and as only three are used for bonding the other one is delocalised.

The layers in graphite are able to slide over each other because there are only weak intermolecular forces holding them together. This is why graphite is soft and slippery. These properties make graphite suitable for use as a lubricant.

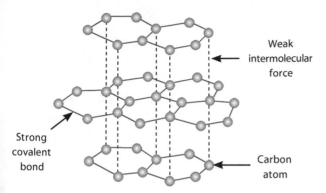

Weak intermolecular force ←

Strong covalent bond →

← Carbon atom

Like diamond, there are lots of strong covalent bonds in graphite so it has a high melting point.

The delocalised electrons allow graphite to conduct electricity and heat.

Graphene and fullerenes

Graphene is a single layer of graphite and so it is one atom thick.

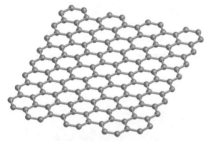

Fullerenes are molecules made up of carbon atoms and they have hollow shapes. The structure of fullerenes is based on hexagonal rings of carbon atoms but they may also contain rings with five or seven carbon atoms.

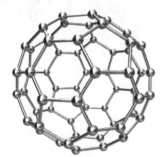

buckminsterfullerene (C_{60}) was the first fullerene to be discovered

Carbon nanotubes are cylindrical fullerenes.

Fullerenes have high…

● tensile strength

● electrical conductivity

● thermal conductivity.

Fullerenes can be used…

● for drug delivery into the body

● as lubricants

● for reinforcing materials, e.g. tennis rackets.

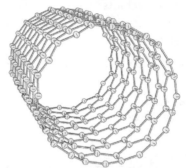

SUMMARY

● **Most metals we use are alloys; alloys are stronger than pure metals.**

● **Diamond and graphite have giant covalent structures; they both have high melting points.**

QUESTIONS

QUICK TEST

1. Why do metals generally have high melting points?

2. Suggest two uses of fullerenes.

EXAM PRACTICE

1. Magnalium is an alloy of aluminium and magnesium. It is used in aircraft manufacture because it is both strong and low in density.

   a) What is an alloy? **[1 mark]**

   b) Explain why magnalium is stronger than pure aluminium. **[2 marks]**

2. Carbon atoms can arrange themselves in different ways. One of those ways is in a form known as graphite.

   a) Describe the structure of graphite. **[3 marks]**

   b) Explain why graphite can conduct electricity. **[2 marks]**

Mass and equations

Conservation of mass

The total mass of reactants in a chemical reaction is equal to the total mass of the products because atoms are not created or destroyed.

Chemical reactions are represented by balanced symbol equations.

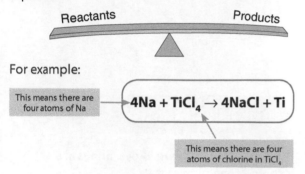

Reactants   Products

For example:

This means there are four atoms of Na

$$4Na + TiCl_4 \rightarrow 4NaCl + Ti$$

This means there are four atoms of chlorine in $TiCl_4$

Relative formula mass

The relative formula mass (M_r) of a compound is the sum of the relative atomic masses (see the periodic table at the back of this book) of the atoms in the formula.

For example:

● The M_r of MgO is 40 $(24 + 16)$

● The M_r of H_2SO_4 is 98 $[(2 \times 1) + 32 + (4 \times 16)]$

In a balanced symbol equation, the sum of the relative formula masses of the reactants equals the sum of the relative formula masses of the products.

Example	$CaCO_3 + 2HCl \rightarrow CaCl_2 + H_2O + CO_2$
Sum of relative formula masses:	$100 + (2 \times 36.5) = 173$ $111 + 18 + 44 = 173$

Mass changes when a reactant or product is a gas

During some chemical reactions, there can appear to be a change in mass. When copper is heated its mass actually increases because oxygen is being added to it.

copper + oxygen → copper oxide

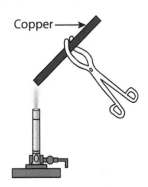

Copper

The mass of copper oxide formed is equal to the starting mass of copper plus the mass of the oxygen that has been added to it.

During a thermal decomposition reaction of a metal carbonate, the final mass of remaining metal oxide solid is less than the starting mass. This is because when the metal carbonate thermally decomposes it releases carbon dioxide gas into the atmosphere.

copper carbonate → copper oxide + carbon dioxide

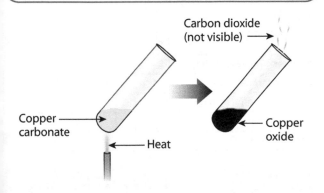

Carbon dioxide (not visible)

Copper carbonate

Copper oxide

Heat

Example

Starting mass of copper carbonate = 8.00 g
Final mass of copper oxide = 5.15 g

Therefore, mass of carbon dioxide released to the atmosphere = 2.85 g
(8.00 g – 5.15 g)

SUMMARY

● Balanced equations are used to show chemical reactions.

● Relative formula mass of a compound is the sum of the relative atomic masses of the atoms in the formula.

● Mass changes when a reactant or product is a gas.

QUESTIONS

QUICK TEST

1. In a chemical reaction the total mass of reactants is 9.22 g. Will the expected mass of all the products be lower, higher or the same as 9.22 g?

2. What is relative formula mass?

3. When 6.3 g of zinc carbonate is heated will the mass of remaining solid be lower, higher or the same as 6.3 g?

4. With reference to the periodic table at the back of this book, work out the relative formula mass of the compound $Mg(OH)_2$.

EXAM PRACTICE

1. Magnesium oxide can be made by heating magnesium in air or by heating magnesium carbonate.

   The equation for the reaction that occurs when magnesium carbonate is heated is shown below.

   $$MgCO_{3\,(s)} \rightarrow MgO_{(s)} + CO_{2\,(g)}$$

   a) What is the relative formula mass of magnesium carbonate? **[1 mark]**

   b) If 10g of magnesium carbonate formed 4.76g of magnesium oxide upon heating, determine the mass of carbon dioxide produced in this experiment. **[1 mark]**

   c) What name is given to the type of reaction occurring in part b? **[1 mark]**

   d) What will happen to the mass of magnesium metal when it is heated in air? Explain your answer. **[2 marks]**

Moles, masses, empirical and molecular formula

HT Moles and Avogadro's constant

Amounts of chemicals are measured in moles (mol). The number of atoms, molecules or ions in a mole of a given substance is 6.02×10^{23}. This value is known as Avogadro's constant.

Avogadro's constant = 602 000 000 000 000 000 000 000

For example:

● 1 mole of carbon contains 6.02×10^{23} carbon atoms

● 1 mole of sulfur dioxide (SO_2) contains 6.02×10^{23} sulfur dioxide molecules.

Moles and relative formula mass

The mass of one mole of a substance in grams is equal to its relative formula mass (M_r).

For example:

● The mass of one mole of carbon is 12 g.

● The mass of one mole of sulfur dioxide is 64 g.

The number of moles can be calculated using the following formula:

$$moles = \frac{mass}{M_r}$$

For example, the number of moles of carbon in $48\,g = \frac{48}{12} = 4$

By rearranging the above equation, the relative formula mass of a compound can be worked out from the number of moles and mass.

Example

Calculate the relative formula mass of the compound given that 0.23 moles has a mass of 36.8 g.

$$M_r = \frac{mass}{moles} \qquad M_r = \frac{36.8}{0.23} = 160$$

Amounts of substances in equations

Balanced symbol equations give information about the number of moles of reactants and products. For example:

$$2Mg + O_2 \rightarrow 2MgO$$

This equation tells us that 2 moles of magnesium react with one mole of oxygen to form 2 moles of magnesium oxide.

This means that 48 g of Mg (the mass of 2 moles of Mg) reacts with 32 g of oxygen to form 80 g of magnesium oxide.

	2Mg	+	O_2	→	2MgO
Number of moles reacting	2		1		2
Relative formula mass	24		32		40
Mass reacting/formed (g)	48		32		80

We can use this relationship between mass and moles to calculate reacting masses.

Example: Calculate the mass of magnesium oxide formed when 12 g of magnesium reacts with an excess of oxygen.

	2Mg	+	O_2	→	2MgO
Number of moles reacting	2		1		2
Relative formula mass	24		32		40
Mass reacting/formed (g)	48		32		80
Reacting mass (g)	12				

To get from 48 to 12 we divide by 4

Therefore to find the mass of magnesium oxide formed we divide 80 by 40

The mass of magnesium oxide formed is therefore 20 g.

Empirical formula

The empirical formula is the simplest whole number ratio of each type of atom present in a compound. For example, hexene (C_6H_{12}) has the empirical formula CH_2.

You can work out the empirical formula of a substance from its chemical formula. For example, the empirical formula of ethanoic acid (CH_3COOH) is CH_2O.

The empirical formula of a compound can be calculated from either:

● the percentage composition of the compound by mass

or

● the mass of each element in the compound.

To calculate the empirical formula:

1 List all the elements in a compound:

2 Divide the data for each element by the relative atomic mass (A_r) of the element (to find the number of moles).

3 Select the smallest answer from step 2 and divide each answer by that result to obtain a ratio.

4 The ratio may need to be scaled up to give whole numbers.

Example 1

What is the empirical formula of a hydrocarbon containing 75% carbon? (Hydrogen = 25%)

1   Carbon   :   Hydrogen

2 $\frac{75}{12}$: $\frac{25}{1}$

      6.25   :   25

3 ÷ 6.25   :   ÷ 6.25

4   1   :   4

So the empirical formula is CH_4.

Example 2

What is the empirical formula of a compound containing 24g of carbon, 8g of hydrogen and 32g of oxygen?

1   Carbon : Hydrogen Oxygen

2 $\frac{24}{12}$: $\frac{8}{1}$ $\frac{32}{16}$

    2 : 8   2

3 ÷ 2 : ÷ 2   ÷ 2

4   1 : 4   1

So the empirical formula is CH_4O.

Molecular formula

The molecular formula is the actual whole number ratio of each type of atom in a compound. It can be the same as the empirical formula or a multiple of the empirical formula. To convert an empirical formula into a molecular formula, you also need to know the relative formula mass of the compound.

Example: A compound has an empirical formula of CH_2 and an M_r of 42. What is its molecular formula?

(A_r for C = 12 and A_r for H = 1)

Work out the relative formula mass of the empirical formula	$= 12 + (2 \times 1) = 14$
Then divide the actual M_r by the empirical formula M_r	$= \frac{42}{14}$
This gives the multiple.	$= 3$

The molecular formula is C_3H_6.

SUMMARY

● **Amounts of chemicals are measured in moles.**

● **The mass of one mole of a substance in grams is equal to its relative formula mass (M_r).**

● **Empirical formula is the simplest whole number ratio of each type of atom present in a compound.**

QUESTIONS

QUICK TEST

1. Calculate the empirical formula of a compound containing 0.35 g of lithium and 0.40 g of oxygen.

2. An oxide of phosphorus has the empirical formula P_2O_5 and M_r of 284. What is its molecular formula?

EXAM PRACTICE

HT 1. The equation for the reaction between aluminium and oxygen is shown below.

$4Al_{(s)} + 3O_{2\,(g)} \rightarrow 2Al_2O_{3\,(s)}$

Calculate the mass of aluminium oxide formed when 4.32g of aluminium reacts with oxygen. **[3 marks]**

HT Moles, solutions and equations

Concentration of solutions in g/dm³

Many chemical reactions take place in solutions. The concentration of a solution can be measured in mass of solute per given volume of solution, e.g. grams per dm³ ($1 dm^3 = 1000 cm^3$).

For example, a solution of 5 g/dm³ has 5 g of solute dissolved in 1 dm³ of water. It has half the concentration of a 10 g/dm³ solution of the same solute.

The mass of solute in a solution can be calculated if the concentration and volume of solution are known.

> **Example**
>
> Calculate the mass of solute in 250 cm³ of a solution whose concentration is 8 g/dm³.
>
> **Step 1:** Divide the mass by 1000 (this gives you the mass of solute in 1 cm³).
>
> $8 \div 1000 = 0.008$ g/cm³
>
> **Step 2:** Multiply this value by the volume specified.
>
> $0.008 \times 250 = 2$ g

Using moles to balance equations

The masses of reactants / products in an equation and the M_r values can be used to work out the balancing numbers in a symbol equation.

Example

Balance the equation below given that 8 g of CH_4 reacts with 32 g of oxygen to form 22 g of CO_2 and 18 g of H_2O

$$...CH_4 + ...O_2 \rightarrow ...CO_2 + ...H_2O$$

Chemical	CH_4	O_2	CO_2	H_2O
Mass (from question)	8	32	22	18
M_r	16	32	44	18
Moles = $\dfrac{mass}{M_r}$	$\dfrac{8}{16} = 0.5$	$\dfrac{32}{32} = 1$	$\dfrac{22}{44} = 0.5$	$\dfrac{18}{18} = 1$

We can make this a whole number ratio by dividing all answers by the smallest answer

÷ 0.5		$\dfrac{0.5}{0.5} = 1$	$\dfrac{1}{0.5} = 2$	$\dfrac{0.5}{0.5} = 1$	$\dfrac{1}{0.5} = 2$

The balanced equation is therefore:

$$......CH_4 + ...O_2 \rightarrow ......CO_2 + ...H_2O$$

SUMMARY

● Concentration of a solution can be measured in mass of solute per given volume of solution.

● A solute is a solid that dissolves in a liquid to form a solution.

● Moles can be used to work out the balancing numbers in a symbol equation.

QUESTIONS

QUICK TEST

1. Calculate the mass of solute in 140 cm^3 of a solution whose concentration is 6 g/dm^3.

2. Balance the equation below for the reaction that occurs when 7 g of silicon reacts with 35.5 g of chlorine to form 42.5 g of silicon chloride.

$$..........Si + ...Cl_2 \rightarrow ....SiCl_4$$

3. What is a solute?

EXAM PRACTICE

1. Which of the following sodium hydroxide solutions has the highest concentration in g/dm^3?

   Solution A contains 27 g of solute in 820 cm^3

   Solution B contains 19g of solute in 560 cm^3

   Show your working.   **[3 marks]**

2. An oxide of iron, X, has a relative formula mass of 232. In an experiment 16.8g of iron reacts with 6.4g of oxygen to make 23.2g of X.

   Use this information to write a balanced symbol equation for the reaction of iron.

   Use X in your equation to represent the oxide of iron.   **[3 marks]**

Reactivity of metals and metal extraction

Reaction of metals with oxygen

Many metals react with oxygen to form metal oxides, for example:

> **copper + oxygen → copper oxide**

These reactions are called oxidation reactions. Oxidation reactions take place when a chemical gains oxygen.

When a substance loses oxygen it is called a reduction reaction.

The reactivity series

When metals react they form positive ions. The more easily the metal forms a positive ion the more reactive the metal.

Calcium and magnesium are both in group 2 of the periodic table so will form 2+ ions when they react. Calcium is more reactive than magnesium so it has a greater tendency/is more likely to form the 2+ ion.

Decreasing reactivity ↓

Metal	Reaction with water	Reaction with acid
Potassium	very vigorous	explosive
Sodium	vigorous	dangerous
lithium	steady	very vigorous
calcium	steady fizzing and bubbling	vigorous
magnesium	slow reaction	steady fizzing and bubbling
aluminium	slow reaction	steady fizzing and bubbling
*Carbon		
zinc	very slow reaction	gentle fizzing and bubbling
iron	extremely slow	slight fizzing and bubbling
*Hydrogen		
copper	no reaction	no reaction
silver	no reaction	no reaction
gold	no reaction	no reaction

* included for comparison (because these non-metals can displace metals)

The reactivity series can also be used to predict displacement reactions.

> **zinc + copper oxide → zinc oxide + copper**

In this reaction…

- zinc displaces (i.e. takes the place of) copper
- zinc is oxidised (i.e. it gains oxygen)
- copper oxide is reduced (i.e. it loses oxygen).

HT Oxidation and reduction in terms of electrons

Oxidation and reduction can also be defined in terms of electrons:

O oxidation

I is

L loss (of electrons)

R reduction

I is

G gain (of electrons)

This mnemonic can be useful to work out what is being oxidised and reduced in displacement reactions, for example:

magnesium + copper sulfate → magnesium sulfate + copper

$Mg + CuSO_4 \rightarrow MgSO_4 + Cu$

The ionic equation for this reaction is:

Mg **loses electrons** to become Mg^{2+}

Oxidation

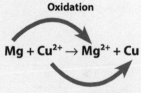

$$Mg + Cu^{2+} \rightarrow Mg^{2+} + Cu$$

Reduction

Cu^{2+} **gains electrons** to become Cu

Extraction of metals and reduction

Unreactive metals, such as gold, are found in the Earth's crust as pure metals. Most metals are found as compounds and chemical reactions are required to extract the metal. The method of extraction depends on the position of the metal in the reactivity series.

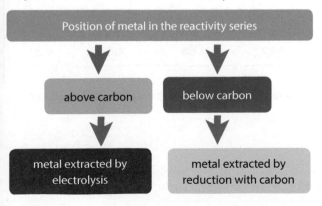

For example, iron is found in the earth as iron(III) oxide, Fe_2O_3. The iron(III) oxide can be reduced by reacting it with carbon.

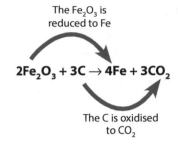

The Fe_2O_3 is reduced to Fe

$$2Fe_2O_3 + 3C \rightarrow 4Fe + 3CO_2$$

The C is oxidised to CO_2

SUMMARY

- **An oxidation reaction is when a metal reacts with oxygen. A reduction reaction is when a substance loses oxygen.**
- **The reactivity series can be used to predict displacement reactions.**

QUESTIONS

QUICK TEST

1. Suggest two metals from the reactivity series that are extracted by reduction with carbon.

2. Which metal is the most reactive – magnesium or iron?

3. Write a word equation for the oxidation of aluminium to form aluminium oxide.

4. Both sodium and lithium react to form 1+ ions. Which one of these metals is more likely to form this ion?

EXAM PRACTICE

1. Magnesium reacts with copper(II) oxide to form magnesium oxide and copper.

   a) Explain why this reaction takes place. **[1 mark]**

   b) In this reaction which substance has been reduced?

   Explain your answer. **[2 marks]**

Reactions of acids

Reactions of acids with metals

Acids react with metals that are above hydrogen in the reactivity series to make salts and hydrogen, for example:

$$\text{magnesium} + \text{hydrochloric acid} \rightarrow \text{magnesium chloride} + \text{hydrogen}$$

This is a salt

HT The reactions of metals with acids are redox reactions. The ionic equation for the reaction of magnesium with hydrochloric acid is:

$$Mg + 2H^+ \rightarrow Mg^{2+} + H_2$$

The metal (in this case magnesium) is oxidised, i.e. it loses electrons.

The hydrogen ions are reduced, i.e. they gain electrons.

Mg is **oxidised**, i.e. it loses electrons (to form Mg^{2+})

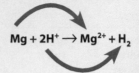

$$Mg + 2H^+ \rightarrow Mg^{2+} + H_2$$

H^+ is **reduced**, i.e. it gains electrons (to form H_2)

Neutralisation of acids and the preparation of salts

Acids can be neutralised by the following reactions.

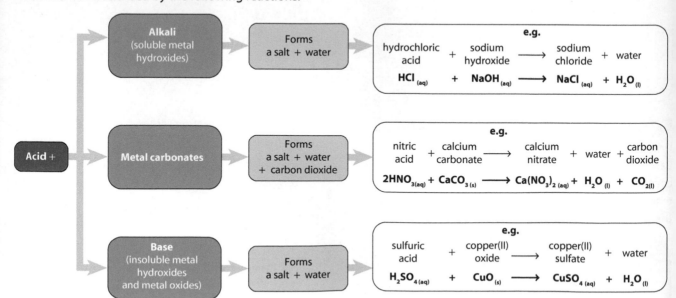

Acid +

Alkali (soluble metal hydroxides) → Forms a salt + water → e.g.

hydrochloric acid + sodium hydroxide ⟶ sodium chloride + water

$$HCl_{(aq)} + NaOH_{(aq)} \longrightarrow NaCl_{(aq)} + H_2O_{(l)}$$

Metal carbonates → Forms a salt + water + carbon dioxide → e.g.

nitric acid + calcium carbonate ⟶ calcium nitrate + water + carbon dioxide

$$2HNO_{3(aq)} + CaCO_{3(s)} \longrightarrow Ca(NO_3)_{2(aq)} + H_2O_{(l)} + CO_{2(l)}$$

Base (insoluble metal hydroxides and metal oxides) → Forms a salt + water → e.g.

sulfuric acid + copper(II) oxide ⟶ copper(II) sulfate + water

$$H_2SO_{4(aq)} + CuO_{(s)} \longrightarrow CuSO_{4(aq)} + H_2O_{(l)}$$

The first part of the salt formed contains the positive ion (usually the metal) from the alkali, base or carbonate followed by...

● chloride if hydrochloric acid was used

● sulfate if sulfuric acid was used

● nitrate if nitric acid was used.

For example, when calcium hydroxide is reacted with sulfuric acid, the salt formed is calcium sulfate.

Making salts

Salts can be either soluble or insoluble. The majority of salts are soluble.

The general rules for deciding whether a salt will be soluble are as follows.

- All common sodium, potassium and ammonium salts are soluble.
- All nitrates are soluble.
- All common chlorides, except silver chloride, are soluble.
- All common sulfates, except barium and calcium, are soluble.
- All common carbonates are insoluble, except potassium, sodium and ammonium.

(ws) Preparation of soluble salts

Soluble salts can be prepared by the following method.

For example, copper(II) sulfate crystals can be made by reacting copper(II) oxide with sulfuric acid.

Add copper(II) oxide to sulfuric acid and stir

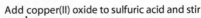

Sulfuric acid

Copper(II) oxide to sulfuric acid → Filter to remove any unreacted copper(II) oxide → Evaporate to leave behind blue crystals of the 'salt' copper(II) sulfate

Copper(II) sulfate

SUMMARY

- Acids can be neutralised by reacting the acid with an alkali, a metal carbonate or a base.
- Salts can be either soluble or insoluble. Most salts are soluble.
- Soluble salts can be made by reacting acid with a solid, filtering off the excess solid and then obtaining the salt by crystallisation.

QUESTIONS

QUICK TEST

1. What is crystallisation?

(HT) 2. Identify the species that is oxidised in the following reaction.

$$Fe + 2H^+ \rightarrow Fe^{2+} + H_2$$

3. Write a word equation for the reaction that occurs when aluminium reacts with sulfuric acid.

4. Name the salt formed when zinc oxide reacts with nitric acid.

5. Which one of the following salts is insoluble?
   sodium carbonate
   calcium sulfate
   copper(II) nitrate

EXAM PRACTICE

1. Magnesium oxide powder reacts with hydrochloric acid to form water and the soluble salt magnesium chloride.

   a) Write a balanced symbol equation for this reaction.

      Include state symbols. **[2 marks]**

   b) Describe how a pure, dry sample of magnesium chloride can be prepared from magnesium oxide and hydrochloric acid.

      Include any necessary equipment in your answer. **[5 marks]**

pH, neutralisation, acid strength and electrolysis

Indicators, the pH scale and neutralisation reactions

Indicators are useful dyes that become different colours in acids and alkalis.

Indicator	Colour in acid	Colour in alkali
Litmus	Red	Blue
Phenolphthalein	Colourless	Pink
Methyl orange	Pink	Yellow

The pH scale measures the acidity or alkalinity of a solution. The pH scale runs from 0 to 14 and the pH of a solution can be measured using universal indicator or a pH probe.

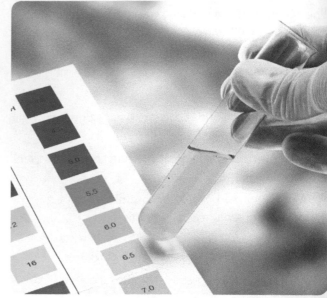

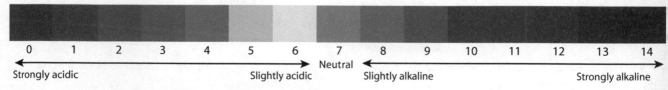

| 0 | 1 | 2 | 3 | 4 | 5 | 6 | 7 | 8 | 9 | 10 | 11 | 12 | 13 | 14 |

Neutral

Strongly acidic — Slightly acidic — Slightly alkaline — Strongly alkaline

Acids are solutions that contain hydrogen ions (H^+). The higher the concentration of hydrogen ions, the more acidic the solution (i.e. the lower the pH).

Alkalis are solutions that contain hydroxide ions (OH^-). The higher the concentration of hydroxide ions, the more alkaline the solution (i.e. the higher the pH).

When an acid is neutralised by an alkali, the hydrogen ions from the acid react with the hydroxide ions in the alkali to form water.

$$H^+_{(aq)} + OH^-_{(aq)} \rightarrow H_2O_{(l)}$$

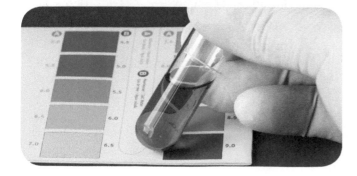

HT Strong acids, such as hydrochloric, sulfuric and nitric acids, are those that completely ionise in aqueous solution. For example:

$$HCl_{(aq)} \rightarrow H^+_{(aq)} + Cl^-_{(aq)}$$

This means dissolved in water

Weak acids, such as ethanoic, citric and carbonic acids, only partially ionise in water. For example:

$$CH_3COOH_{(aq)} \rightleftharpoons CH_3COO^-_{(aq)} + H^+_{(aq)}$$

This sign means that the reaction is reversible, i.e. that the acid does not fully ionise

The pH of an acid is a measure of the concentration of hydrogen ions. When two different acids of the same concentration have different pH values the strongest acid will have the lowest pH.

The pH scale is a logarithmic scale. As the pH decreases by 1 unit (e.g. from 3 to 2) the hydrogen ion concentration increases by a factor of 10.

Electrolysis

An electric current is the flow of electrons through a conductor but it can also flow by the movement of ions through a solution or a liquid.

Covalent compounds do not contain free electrons or ions that can move. So they will not conduct electricity when solid, liquid, gas or in solution.

The ions in:

- an **ionic solid** are fixed and cannot move
- an **ionic substance** that is **molten** or in **solution** are free to move.

Electrolysis is a chemical reaction that involves passing electricity through an electrolyte. An electrolyte is a liquid that conducts electricity. Electrolytes are either molten ionic compounds or solutions of ionic compounds. Electrolytes are decomposed during electrolysis.

- The positive ions (**cations**) move to, and discharge at, the negative electrode (**cathode**).
- The negative ions (**anions**) move to, and discharge at, the positive electrode (**anode**).

Electrons are removed from the anions at the anode. These electrons then flow around the circuit to the cathode and are transferred to the cations.

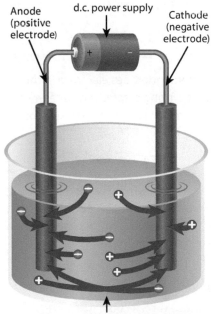

Anode (positive electrode)   d.c. power supply   Cathode (negative electrode)

Electrolyte (liquid that conducts electricity and decomposes in electrolysis)

SUMMARY

- The pH scale measures the acidity or alkalinity of a solution.
- Acids are solutions that contain hydrogen ions; alkalis are solutions that contain hydroxide ions.
- Electrolysis is a chemical reaction that involves passing electricity through an electrolyte.

QUESTIONS

QUICK TEST

1. What is an electrolyte?
2. Which ion is responsible for solutions being acidic?
3. What type of substance has a pH of more than 7?
HT **4.** What is the difference between a strong acid and a weak acid?
5. What name is given to positive ions?

EXAM PRACTICE

1. Sodium hydroxide is a common laboratory alkali. In an experiment, hydrochloric acid was added to sodium hydroxide solution containing phenolphthalein indicator until the indicator changed colour.

   a) What colour will the phenolphthalein be at the beginning and end of the experiment?   **[2 marks]**

   b) Name the ion that causes solutions to be alkaline.   **[1 mark]**

   c) Write the ionic equation for the reaction between sodium hydroxide and hydrochloric acid.   **[1 mark]**

   HT d) Hydrochloric acid is a strong acid.

   Explain the meaning of the term 'strong acid'.   **[2 marks]**

Applications of electrolysis

Electrolysis of molten ionic compounds

When an ionic compound melts, electrostatic forces between the charged ions in the crystal lattice are broken down, meaning that the ions are free to move.

When a direct current is passed through a molten ionic compound:

- positively charged ions are attracted towards the **negative electrode** (cathode)
- negatively charged ions are attracted towards the **positive electrode** (anode).

For example, in the electrolysis of molten lead bromide:

- positively charged lead ions are attracted towards the cathode, forming lead
- negatively charged bromide ions are attracted towards the anode, forming bromine.

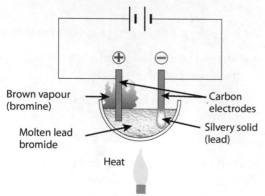

Brown vapour (bromine)

Carbon electrodes

Molten lead bromide

Silvery solid (lead)

Heat

When ions get to the oppositely charged electrode they are **discharged**, i.e. they lose their charge. For example, in the electrolysis of molten lead bromide, the non-metal ion loses electrons to the positive electrode to form a bromine atom. The bromine atom then bonds with a second atom to form a bromine molecule.

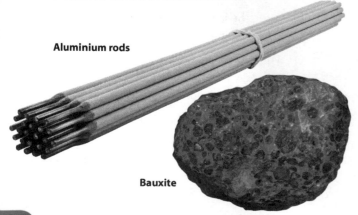

Aluminium rods

Bauxite

ⓌⓈ Using electrolysis to extract metals

Aluminium is the most abundant metal in the Earth's crust. It must be obtained from its ore by electrolysis because it is too reactive to be extracted by heating with carbon. The electrodes are made of graphite (a type of carbon). The aluminium ore (bauxite) is purified to leave aluminium oxide, which is then melted so that the ions can move. Cryolite is added to increase the conductivity and lower the melting point.

When a current passes through the molten mixture:

- positively charged aluminium ions move towards the negative electrode (**cathode**) and form aluminium
- negatively charged oxygen ions move towards the positive electrode (**anode**) and form oxygen.

The positive electrodes gradually wear away (because the graphite electrodes react with the oxygen to form carbon dioxide gas). This means they have to be replaced every so often. Extracting aluminium can be quite an expensive process because of the cost of the large amounts of electrical energy needed to carry it out.

Electrolysis of aqueous solutions

When a solution undergoes electrolysis, there is also water present. During electrolysis water molecules break down into hydrogen ions and hydroxide ions.

$$H_2O_{(l)} \rightarrow H^+_{(aq)} + OH^-_{(aq)}$$

This means that when an aqueous compound is electrolysed there are two cations present (H^+ from water and the metal cation from the compound) and two anions present (OH^- from water and the anion from the compound).

For example, in copper(II) sulfate solution…

- cations present: Cu^{2+} and H^+
- anions present: SO_4^{2-} and OH^-

At the positive electrode (anode): *Oxygen is produced unless the solution contains halide ions*. In this case the oxygen is produced.

At the negative electrode (cathode): *The least reactive element is formed*. The reactivity series on page 34 will be helpful here. In this case, hydrogen is formed.

Example

What are the three products of the electrolysis of sodium chloride solution?

Cations present: Na^+ and H^+

Q. What happens at the cathode?

A. Hydrogen is less reactive than sodium therefore hydrogen gas is formed.

Anions present: Cl^- and OH^-

Q. What happens at the anode?

A. A halide ion (Cl^-) is present therefore chlorine will be formed.

The sodium ions and hydroxide ions stay in solution (i.e. sodium hydroxide solution remains).

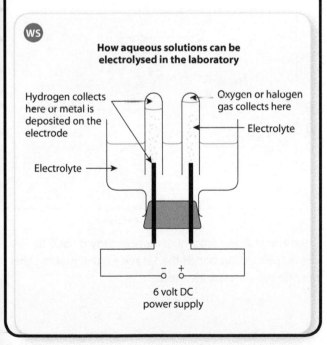

WS

How aqueous solutions can be electrolysed in the laboratory

Hydrogen collects here or metal is deposited on the electrode

Oxygen or halogen gas collects here

Electrolyte

Electrolyte

6 volt DC power supply

Solution	Product at cathode	Product at anode
copper chloride	copper	chlorine
sodium sulfate	hydrogen	oxygen
water (diluted with sulfuric acid to aid conductivity)	hydrogen	oxygen

HT Half-equations

During electrolysis, the cation that is discharged at the cathode gains electrons (is reduced) to form the element. For example:

$$Cu^{2+} + 2e^- \rightarrow Cu$$

$$2H^+ + 2e^- \rightarrow H_2$$

At the anode the anion loses electrons (is oxidised). For example:

$$2Cl^- \rightarrow Cl_2 + 2e^-$$

this can also be written as
$$2Cl^- - 2e^- \rightarrow Cl_2$$

$$4OH^- \rightarrow O_2 + 2H_2O + 4e^-$$

or $4OH^- - 4e^- \rightarrow O_2 + 2H_2O$

SUMMARY

● When a direct current is passed through a molten ionic compound, positively charged ions move towards the cathode, and negatively charged ions move towards the anode.

● Electrolysis can be used to obtain aluminium from its ore.

● During electrolysis, water molecules break down into hydrogen ions and hydroxide ions.

QUESTIONS

QUICK TEST

1. When molten copper chloride is electrolysed, what will be formed at the cathode and anode?

2. What ions do water molecules break down into during electrolysis?

EXAM PRACTICE

1. Name the products and state at which electrode they will be formed when copper sulfate solution is electrolysed. **[2 marks]**

Energy changes in reactions

Exothermic and endothermic reactions

Energy is not created or destroyed during chemical reactions, i.e. the amount of energy in the universe at the end of a chemical reaction is the same as before the reaction takes place.

Type of reaction	Is energy given out or taken in?	What happens to the temperature of the surroundings?
Exothermic	out	increases
Endothermic	in	decreases

Examples of exothermic reactions include…

- combustion
- neutralisation
- many oxidation reactions
- precipitation reactions
- displacement reactions.

Everyday applications of exothermic reactions include self-heating cans and hand warmers.

Examples of endothermic reactions include…

- thermal decomposition
- the reaction between citric acid and sodium hydrogen carbonate.

Some changes, such as dissolving salts in water, can be either exothermic or endothermic. Some sports injury packs are based on endothermic reactions.

Reaction profiles

For a chemical reaction to occur, the reacting particles must collide together with sufficient energy. The minimum amount of energy that the particles must have in order to react is known as the 'activation energy'.

Reaction profiles can be used to show the relative energies of reactants and products, the activation energy and the overall energy change of a reaction.

Reaction profile for an exothermic reaction

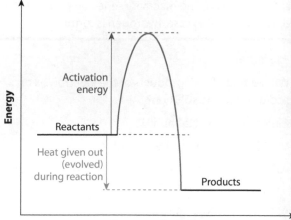

Chemical reactions in which more energy is made when new bonds are made than was used to break the existing bonds are **exothermic**.

Reaction profile for an endothermic reaction

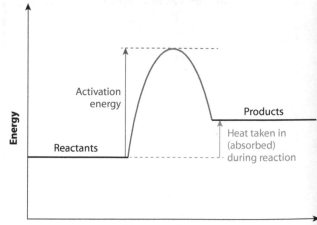

Chemical reactions in which more energy is used to break the existing bonds than is released in making the new bonds are **endothermic**.

HT The energy change of reactions

During a chemical reaction:

- bonds are broken in the reactant molecules – this is an endothermic process
- bonds are made to form the product molecules – this is an exothermic process.

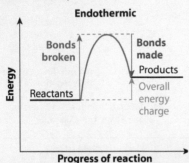

Endothermic

If more energy is used to break the bonds than is released when the bonds are made, then the reaction is endothermic.

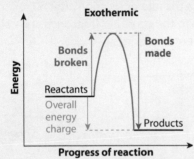

Exothermic

If more energy is released when bonds are made than is used to break the bonds, then the reaction is exothermic.

Example

Hydrogen is burned in oxygen to produce water:

hydrogen + oxygen → water

$$2H_{2(g)} + O_{2(g)} → 2H_2O_{(l)}$$

$$2H-H + O=O → 2H-O-H$$

The following are **bond energies** for the **reactants** and **products**:

H–H is 436 kJ   O=O is 496 kJ   O–H is 463 kJ

Calculate the energy change.

You can calculate the energy change using this method:

1 Calculate the energy used to break bonds:

$$(2 × H-H) + O=O = (2 × 436) + 496 = \textbf{1368 kJ}$$

2 Calculate the energy released when new bonds are made:

(Water is made up of 2 × O–H bonds.)

$$2 × H-O-H = 2 × (2 × 463) = \textbf{1852 kJ}$$

Enthalpy change (ΔH) = energy used to break bonds – energy released when new bonds are made

$$ΔH = 1368 - 1852$$

$$ΔH = \textbf{-484 kJ}$$

The reaction is **exothermic** because the energy from making the bonds is **more than** the energy needed to break the bonds.

SUMMARY

- **Exothermic reactions give out energy and the temperature of the surroundings increases; endothermic reactions take in energy and the temperature of the surroundings decreases.**
- **For a chemical reaction to occur, reacting particles must collide with sufficient energy known as the activation energy.**

QUESTIONS

QUICK TEST

1. Give one example of an exothermic reaction and one example of an endothermic reaction.
2. During an exothermic reaction, is heat given out or taken in?

QUESTIONS

EXAM PRACTICE

1. Draw and label a reaction profile for an exothermic reaction. **[3 marks]**

HT 2. Calculate the energy change of the following reaction. **[3 marks]**

H–H + Cl–Cl → 2H–Cl

Bond	Energy kJ/mol
H–H	436
Cl–Cl	239
H–Cl	427

Rates of reaction

ⓦⓢ Calculating rates of reactions

The rate of a chemical reaction can be determined by measuring the quantity of a reactant used or (more commonly) the quantity of a product formed over time.

$$\text{mean rate of reaction} = \frac{\text{quantity of reactant used}}{\text{time taken}}$$

$$\text{mean rate of reaction} = \frac{\text{quantity of product formed}}{\text{time taken}}$$

For example, if 46 cm^3 of gas is produced in 23 seconds then the mean rate of reaction is 2 cm^3/s.

ⒽⓉ Rates of reaction can also be determined from graphs.

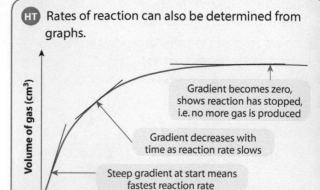

Gradient becomes zero, shows reaction has stopped, i.e. no more gas is produced

Gradient decreases with time as reaction rate slows

Steep gradient at start means fastest reaction rate

If the gradient of the tangent is calculated, this gives a numerical measure of the rate of reaction, e.g. to calculate the rate of reaction after 60 seconds draw a tangent to the curve at 60 seconds and calculate the gradient of this tangent.

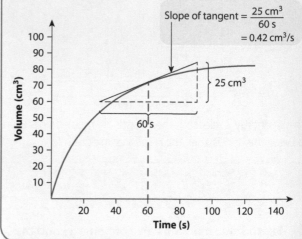

Slope of tangent $= \dfrac{25 \text{ cm}^3}{60 \text{ s}}$
$= 0.42 \text{ cm}^3/\text{s}$

Factors affecting the rates of reactions

There are five factors that affect the rate of chemical reactions:

- concentrations of the reactants in solution
- pressure of reacting gases
- surface area of any solid reactants
- temperature
- presence of a catalyst.

During experiments the rate of a chemical reaction can be found by…

- measuring the mass of the reaction mixture (e.g. if a gas is lost during a reaction)
- measuring the volume of gas produced
- observing a solution becoming opaque or changing colour.

Weighing the reaction mixture

Measuring the volume of gas produced

Observing the formation of a precipitate

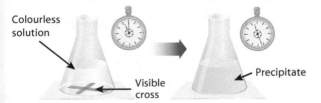

Colourless solution

Visible cross

Precipitate

SUMMARY

- **Rate of a reaction can be determined by measuring the quantity of reactant used or the quantity of product formed.**
- **A number of factors can affect the rate of reaction, including use of a catalyst.**

QUESTIONS

QUICK TEST

1. What is the mean rate of the reaction in which 30 g of reactant is used up over 10 seconds?

2. State two factors that affect the rate of reaction.

EXAM PRACTICE

1. 0.5g of magnesium ribbon was added to a conical flask containing hydrochloric acid of concentration labelled as 100%. A diagram of the apparatus used is shown below.

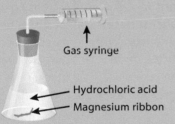

Gas syringe

Hydrochloric acid

Magnesium ribbon

The student found that it took 42 seconds to collect 90 cm³ of hydrogen gas.

a) Calculate the mean rate of this reaction in cm³/s.

Give your answer to 1 decimal place. **[2 marks]**

b) Suggest how the rate of reaction would change if hydrochloric acid of 50% concentration was used. **[1 mark]**

c) Other than changing concentration, suggest one other way that the rate of this reaction can be increased. **[1 mark]**

Collision theory, activation energy and catalysts

Factors affecting rates of reaction and collision theory

Collision theory explains how various factors affect rates of reaction.

It states that, for a chemical reaction to occur...

1 The reactant particles must collide with each other.

AND

2 They must collide with sufficient energy – this amount of energy is known as the activation energy.

Surface area	Temperature	Pressure	Concentration
A smaller particle size means a higher surface area to volume ratio. With smaller particles, more collisions can take place, meaning a greater rate of reaction.	Increasing the temperature increases the rate of reaction because the particles are moving more quickly and so will collide more often. Also, more particles will possess the activation energy, so a greater proportion of collisions will result in a reaction.	At a higher pressure, the gas particles are closer together, so there is a greater chance of them colliding, resulting in a higher rate of reaction.	At a higher concentration, there are more reactant particles in the same volume of solution, which increases the chance of collisions and increases the rate of reaction.
Large pieces – small surface area to volume ratio **Small pieces** – large surface area to volume ratio	**Low temperature** **High temperature**	**Low pressure** **High pressure**	**Low concentration** **High concentration**

Catalysts

Catalysts are chemicals that change the rate of chemical reactions but are not used up during the reaction. Different chemical reactions need different catalysts. In biological systems enzymes act as catalysts.

Catalysts work by providing an alternative reaction pathway of lower activation energy. This can be shown on a reaction profile.

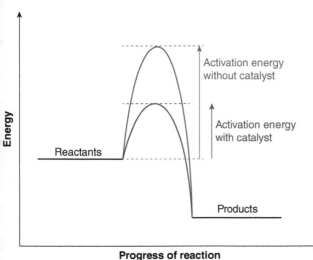

Activation energy without catalyst

Activation energy with catalyst

Reactants

Products

Progress of reaction

Energy

Catalysts are not reactants and so they are not included in the chemical equation.

Enzyme (biological catalyst)

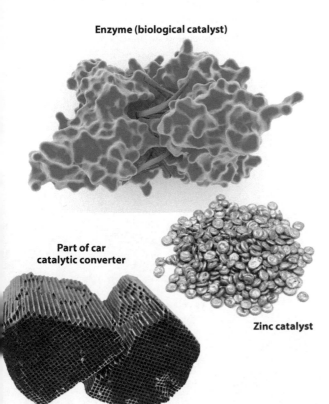

Part of car catalytic converter

Zinc catalyst

SUMMARY

● **Collision theory states that for a chemical reaction to occur, reactant particles must collide with each other and must collide with sufficient energy – activation energy.**

● **Catalysts are chemicals that change the rate of reactions but are not used up during the reaction.**

QUESTIONS

QUICK TEST

1. What is meant by the term 'activation energy'?

2. Why does a lower concentration of solution decrease the rate of a chemical reaction?

3. How do catalysts increase the rate of chemical reactions?

EXAM PRACTICE

1. Sulfur trioxide gas can be made by heating sulfur dioxide gas with oxygen in the presence of a vanadium pentoxide, V_2O_5 catalyst.

   $$2SO_{2\,(g)} + O_{2\,(g)} \rightarrow 2SO_{3\,(g)}$$

   a) Explain why heating the mixture of gases increases the rate of reaction. **[3 marks]**

   b) Explain how the vanadium pentoxide catalyst increases the rate of reaction. **[2 marks]**

   c) Increasing the pressure also increases the rate of reaction. Explain how. **[2 marks]**

Reversible reactions and equilibrium

Reversible reactions

In some reactions, the products of the reaction can react to produce the original reactants. These reactions are called reversible reactions.

For example:

● Heating ammonium chloride:

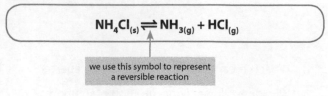

$$NH_4Cl_{(s)} \rightleftharpoons NH_{3(g)} + HCl_{(g)}$$

we use this symbol to represent a reversible reaction

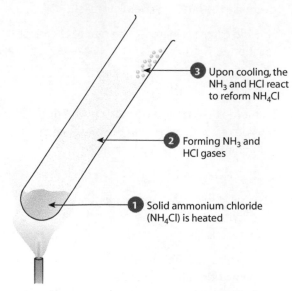

3 Upon cooling, the NH_3 and HCl react to reform NH_4Cl

2 Forming NH_3 and HCl gases

1 Solid ammonium chloride (NH_4Cl) is heated

● Heating hydrated copper(II) sulfate:

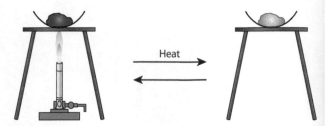

Heat
$$CuSO_4 \cdot 5H_2O_{(s)} \rightleftharpoons CuSO_{4(s)} + 5H_2O_{(l)}$$
blue   white

Heat

If the forward reaction is endothermic (absorbs heat) then the reverse reaction must be exothermic (releases heat).

Equilibrium

When a reversible reaction is carried out in a closed system (nothing enters or leaves) and the rate of the forward reaction is equal to the rate of the reverse reaction, the reaction is said to have reached equilibrium.

HT Equilibrium conditions

The relative amounts of reactants and products at equilibrium depend on the **reaction conditions**. The effect of changing conditions on reactions at equilibrium can be predicted by Le Chatelier's principle, which states that 'for a reversible reaction, if changes are made to the concentration, temperature or pressure (for gaseous reactions) then the system responds to counteract the change'.

Changing concentration

If the concentration of one of the reactants is increased, more products will be formed (to use up the extra reactant) until equilibrium is established again (i.e. the equilibrium moves to the right until a new equilibrium is established). Similarly, if the concentration of one of the products is increased, more reactants will be formed (i.e. the equilibrium moves to the left until a new equilibrium is established).

Changing temperature

Forward reaction	Effect of increasing temperature	Effect of decreasing temperature
Endothermic	Equilibrium moves to right-hand side (i.e. forward reaction)	Equilibrium moves to left-hand side (i.e. reverse reaction)
Exothermic	Equilibrium moves to left-hand side	Equilibrium moves to right-hand side

Changing pressure

In order to predict the effect of changing pressure, the number of molecules of gas on each side of the equation needs to be known:

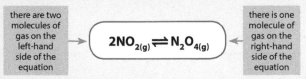

there are two molecules of gas on the left-hand side of the equation → $2NO_{2(g)} \rightleftharpoons N_2O_{4(g)}$ ← there is one molecule of gas on the right-hand side of the equation

If the pressure on a reaction at equilibrium is **increased**, the equilibrium shifts to the side of the equation with the **fewer** molecules of gas. In this case, increasing the pressure will shift the reaction to the right-hand side (i.e. producing more N_2O_4).

SUMMARY

● In reversible reactions, the products of the reaction can react to produce the original reactants.

● When the rate of the forward reaction is equal to the rate of the reverse reaction, the reaction has reached equilibrium.

QUESTIONS

QUICK TEST

1. The forward reaction in a reversible reaction is exothermic. What is the reverse reaction?

HT 2. If the forward reaction is endothermic, what is the effect on equilibrium of increasing temperature?

EXAM PRACTICE

1. Nitrogen and hydrogen gases react to form ammonia, NH_3. This is a reversible reaction.

   a) Write a balanced symbol equation for this reaction. State symbols are not required. **[2 marks]**

   b) When a mixture of these gases is placed in a closed system, eventually a state of equilibrium is reached.

   What is meant by the term equilibrium? **[2 marks]**

HT 2. Methane reacts with steam as shown by the equation below.

$$CH_{4(g)} + H_2O_{(g)} \rightleftharpoons 3H_{2(g)} + CO_{(g)} \quad \Delta H = +206 \text{ kJ/mol}$$

   a) State and explain what will happen to the position of equilibrium if the pressure is increased. **[2 marks]**

   b) State and explain what will happen to the position of equilibrium if the temperature is increased. **[2 marks]**

Crude oil, hydrocarbons and alkanes

Crude oil

This process describes how crude oil is formed.

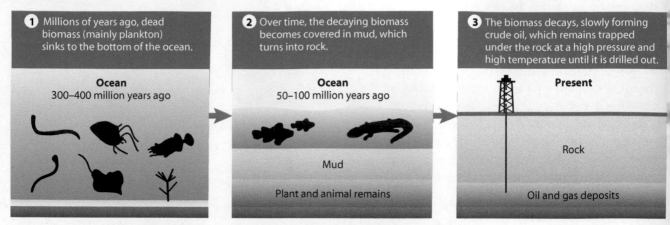

1 Millions of years ago, dead biomass (mainly plankton) sinks to the bottom of the ocean.

Ocean
300–400 million years ago

2 Over time, the decaying biomass becomes covered in mud, which turns into rock.

Ocean
50–100 million years ago

Mud

Plant and animal remains

3 The biomass decays, slowly forming crude oil, which remains trapped under the rock at a high pressure and high temperature until it is drilled out.

Present

Rock

Oil and gas deposits

As it takes so long to form crude oil, we consider it to be a finite or non-renewable resource.

Hydrocarbons and alkanes

Crude oil is a mixture of molecules called hydrocarbons.

Most hydrocarbons are members of a homologous series of molecules called alkanes.

Members of a homologous series…

⦁ have the same general formula

⦁ differ by CH_2 in their molecular formula from neighbouring compounds

⦁ show a gradual trend in physical properties, e.g. boiling point

⦁ have similar chemical properties.

Alkanes are hydrocarbons that have the general formula C_nH_{2n+2}.

Alkane	Methane, CH_4	Ethane, C_2H_6	Propane, C_3H_8	Butane, C_4H_{10}
Displayed formula	H | H—C—H | H	H H | | H—C—C—H | | H H	H H H | | | H—C—C—C—H | | | H H H	H H H H | | | | H—C—C—C—C—H | | | | H H H H

Fractional distillation

Crude oil on its own is relatively useless. It is separated into more useful components (called fractions) by fractional distillation. The larger the molecule, the stronger the intermolecular forces and so the higher the boiling point.

Crude oil is heated until it evaporates.

It then enters a fractionating column which is hotter at the bottom than at the top

where the molecules condense at different temperatures.

Groups of molecules with similar boiling points are collected together. They are called fractions.

The fractions are sent for processing to produce fuels and feedstock (raw materials) for the petrochemical industry

which produces many useful materials, e.g. solvents, lubricants, detergents and polymers (plastics).

Fractionating column

Cool (approximately 25°C)

Refinery gases / LPG (bottled gas)

Petrol (fuel for cars)

Naphtha (making other chemicals)

Kerosene / paraffin (aircraft fuel)

Diesel (fuel for cars / lorries / buses)

Heated crude oil

Fuel oil (fuel for power stations / ships)

Bitumen (tar for roofs and roads)

Hot (approximately 350°C)

Small molecules

Low boiling point

Low viscosity

Burn easily

Large molecules

High boiling point

High viscosity

Don't burn easily

SUMMARY

- Crude oil is a mixture of hydrocarbons.
- Most hydrocarbons are alkanes.
- Crude oil is separated into useful fractions by fractional distillation.

QUESTIONS

QUICK TEST

1. What is crude oil formed from?

2. What is the general formula of alkanes?

3. What is the displayed formula of ethane?

QUESTIONS

EXAM PRACTICE

1. After being extracted from the Earth, crude oil first undergoes fractional distillation.

   a) Explain why crude oil is fractionally distilled. **[1 mark]**

   b) Explain how fractional distillation separates crude oil into different fractions. **[3 marks]**

Combustion and cracking of hydrocarbons

Combustion of hydrocarbons

Some of the fractions of crude oil (e.g. petrol and kerosene) are used as fuels. Burning these fuels releases energy. Combustion (burning) reactions are oxidation reactions. When the fuel is fully combusted, the carbon in the hydrocarbons is oxidised to carbon dioxide and the hydrogen is oxidised to water.

For example, the combustion of methane:

$$CH_{4(g)} + 2O_{2(g)} \longrightarrow CO_{2(g)} + 2H_2O_{(l)}$$

Cracking and alkenes

Many of the long-chain hydrocarbons found in crude oil are not very useful. Cracking is the process of turning a long-chain hydrocarbon into shorter, more useful ones.

Cracking is done by passing hydrocarbon vapour over a hot catalyst or mixing the hydrocarbon vapour with steam before heating it to a very high temperature.

The diagram shows how cracking can be carried out in the laboratory.

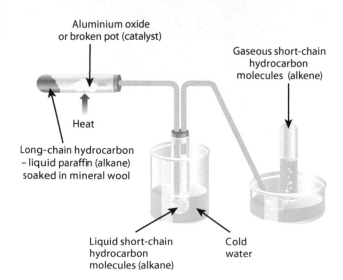

Aluminium oxide or broken pot (catalyst)

Heat

Long-chain hydrocarbon – liquid paraffin (alkane) soaked in mineral wool

Gaseous short-chain hydrocarbon molecules (alkene)

Liquid short-chain hydrocarbon molecules (alkane)

Cold water

Cracking produces alkanes and alkenes. The small-molecule alkanes that are formed during cracking are in high demand as fuels. The alkenes (which are more reactive than alkanes) are mostly used to make plastics by the process of polymerisation and as starting materials for the production of many other chemicals.

Long-chain hydrocarbon

Short-chain hydrocarbons

There are many different equations that can represent cracking. This is because the long hydrocarbon can break in many different places.

A typical equation for the cracking of the hydrocarbon decane ($C_{10}H_{22}$) is:

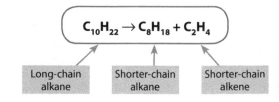

$$C_{10}H_{22} \rightarrow C_8H_{18} + C_2H_4$$

Long-chain alkane | Shorter-chain alkane | Shorter-chain alkene

Example

The cracking of dodecane ($C_{12}H_{26}$) forms one molecule of ethene (C_2H_4), one molecule of butane (C_4H_{10}) and two molecules of another hydrocarbon, as shown in the equation below:

$$C_{12}H_{26} \longrightarrow C_2H_4 + C_4H_{10} + 2 \underline{\hspace{2cm}}$$

Complete the equation for the cracking of dodecane ($C_{12}H_{26}$) by working out the molecular formula of the other hydrocarbon formed.

There are 12 carbon atoms on the left-hand side of the equation and only 6 on the right (2 + 4). Therefore, the two molecules of hydrocarbon must contain 6 carbon atoms in total, i.e. 3 carbon atoms per molecule.

There are 26 hydrogen atoms on the left-hand side of the equation and only 14 on the right (4 + 10). Therefore, the two molecules of hydrocarbon must contain 12 hydrogen atoms in total, i.e. 6 hydrogen atoms per molecule.

Therefore, the molecular formula of the missing hydrocarbon is C_3H_6 and the completed equation is:

$$C_{12}H_{26} \longrightarrow C_2H_4 + C_4H_{10} + 2C_3H_6$$

The presence of alkenes can be detected using bromine water. Alkenes decolourise bromine water but when it is mixed with alkanes, the bromine water stays orange.

Unsaturated alkene (C=C) + bromine water

Saturated alkane (C–C) + bromine water

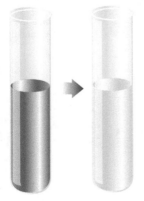

- Complete combustion of hydrocarbons forms carbon dioxide and water.
- Cracking is the process of turning a long-chain hydrocarbon into shorter, more useful molecules.
- Cracking produces alkanes and alkenes.
- The presence of alkenes can be detected using bromine water.

QUESTIONS

QUICK TEST

1. As well as heating with steam, how else can cracking be carried out?

2. What are petrol and kerosene used as?

EXAM PRACTICE

1. When octane, C_8H_{18}, is cracked it can form a molecule of butane and two molecules of an alkene.

   a) Write a balanced symbol equation for this reaction. **[2 marks]**

   b) Describe how the presence of an alkene can be confirmed. **[2 marks]**

Purity, formulations and chromatography

Purity

In everyday language, a pure substance can mean a substance that has had nothing added to it (i.e. in its natural state), such as milk.

In chemistry, a pure substance is a single element or compound (i.e. not mixed with any other substance).

Pure elements and compounds melt and boil at specific temperatures. For example, pure water freezes at 0°C and boils at 100°C. However, if something is added to water (e.g. salt) then the freezing point decreases (i.e. goes below 0°C) and the boiling point rises above 100°C.

Formulations

A formulation is a mixture that has been designed as a useful product. Many formulations are complex mixtures in which each ingredient has a specific purpose.

Formulations are made by mixing the individual components in carefully measured quantities to ensure that the product has the correct properties.

Chromatography

Chromatography is used to separate mixtures of dyes. It is used to help identify substances.

In paper chromatography, a solvent (the mobile phase) moves up the paper (the stationary phase) carrying different components of the mixture different distances, depending on their attraction for the paper and the solvent. In thin layer chromatography (TLC), the stationary phase is a thin layer of an inert substance (e.g. silica) supported on a flat, unreactive surface (e.g. a glass plate).

In the chromatogram below, substance X is being analysed and compared with samples A, B, C, D and E.

It can be seen from the chromatogram that substance X has the same pattern of spots as sample D. This means that sample X and sample D are the same substance. Pure compounds (e.g. compound A) will only produce one spot on a chromatogram.

Fuels

Cleaning materials

Fertilisers

Examples of formulations

Paints

Foods

Medicines

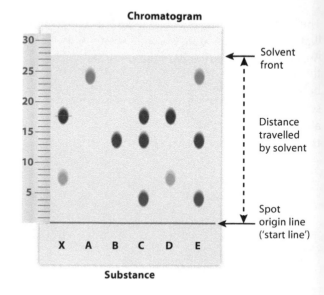

Chromatogram

Solvent front

Distance travelled by solvent

Spot origin line ('start line')

X A B C D E

Substance

The ratio of the distance moved by the compound to the distance moved by the solvent is known as its R_f value.

$$R_f = \frac{\text{distance moved by substance}}{\text{distance moved by solvent}}$$

Different compounds have different R_f values in different solvents. This can be used to help identify unknown compounds by comparing R_f values with known substances.

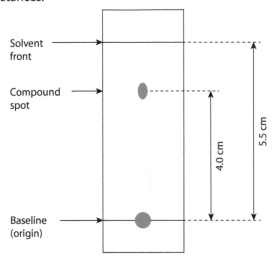

In this case, the R_f value is 0.73 $\left(\frac{4.0}{5.5}\right)$

In **gas chromatography** (GC), the mobile phase is an inert gas (e.g. helium). The stationary phase is a very thin layer of an inert liquid on a solid support, such as beads of silica packed into a long thin tube. GC is a more sensitive method than TLC for separating mixtures, and it allows you to determine the amount of each chemical in the mixture.

SUMMARY

- A pure substance is a single element or compound (i.e. not mixed with any other substance).
- A formulation is a mixture of substances, made to be a useful product.
- Chromatography is a process used to separate mixtures of dyes and help identify substances.

QUESTIONS

QUICK TEST

1. What is chromatography used for?

2. What is a formulation?

3. Give two examples of substances that are formulations.

4. In a chromatogram, the solvent travelled 4 cm and a dye travelled 2.8 cm. What is the R_f value of the dye?

EXAM PRACTICE

1. Paper chromatography was carried out to investigate the dyes present in four different coloured inks. The chromatogram is shown below.

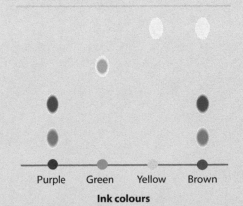

Purple   Green   Yellow   Brown

Ink colours

a) Which two inks only contain one dye?

[1 mark]

b) How can the brown ink be made from the other inks?

[2 marks]

Identification of gases

Testing for hydrogen

Hydrogen burns with a squeaky pop when tested with a lighted splint.

Pop!

Test tube of hydrogen

Lighted splint

Testing for oxygen

Oxygen relights a glowing splint.

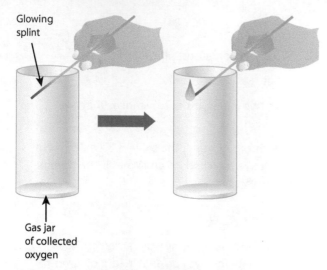

Glowing splint

Gas jar of collected oxygen

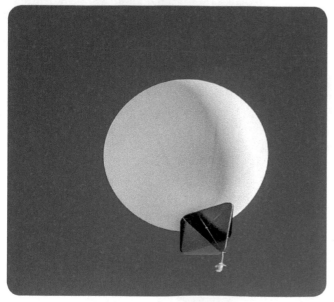

A hydrogen-filled weather balloon

Oxygen is used in breathing masks

Testing for carbon dioxide

When carbon dioxide is mixed with or bubbled through limewater (calcium hydroxide solution) the limewater turns milky (cloudy).

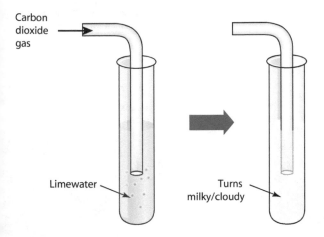

Carbon dioxide gas → Limewater → Turns milky/cloudy

Testing for chlorine

Chlorine turns moist blue litmus paper red before bleaching it and turning it white.

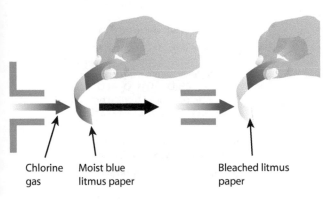

Chlorine gas · Moist blue litmus paper · Bleached litmus paper

Chlorine gas is very toxic

Carbon dioxide is released when fossil fuels are burnt

SUMMARY

- A splint can be used to test for hydrogen and oxygen; test for hydrogen using a lighted splint and test for oxygen using a glowing splint.
- Limewater is used to test for carbon dioxide.
- Litmus paper is used to test for chlorine.

QUESTIONS

QUICK TEST

1. Which gas relights a glowing splint?
2. What is the test for hydrogen gas?
3. What is calcium hydroxide solution also known as?
4. What happens to limewater when it is mixed with carbon dioxide gas?

EXAM PRACTICE

1. When potassium manganate is added to concentrated hydrochloric acid, a gas that is suspected to be chlorine is produced.

   Describe how to confirm that the gas is chlorine. **[3 marks]**

Evolution of the atmosphere

The Earth's early atmosphere

Theories about the composition of Earth's early atmosphere and how the atmosphere was formed have changed over time. Evidence for the early atmosphere is limited because the Earth is approximately 4.6 billion years old and humans were not around to record data.

The table below gives one theory to explain the evolution of the atmosphere.

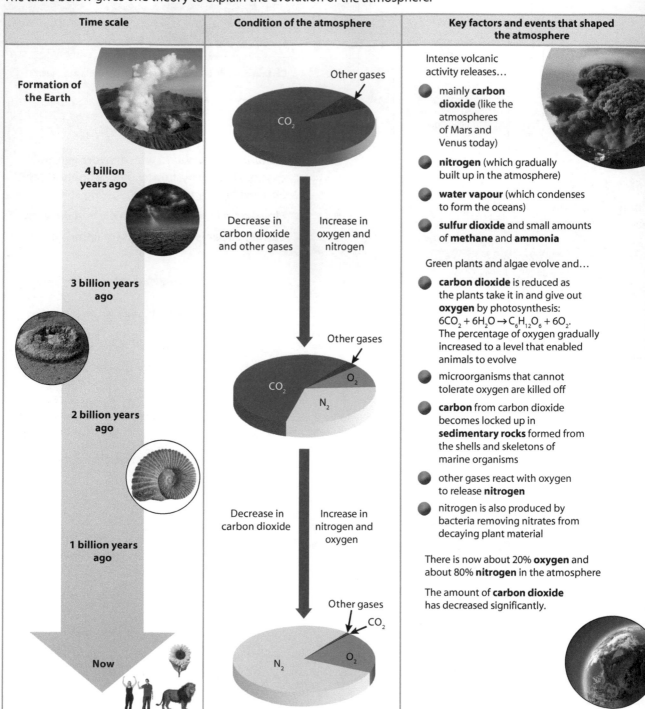

Time scale	Condition of the atmosphere	Key factors and events that shaped the atmosphere
Formation of the Earth	Other gases / CO_2	Intense volcanic activity releases… ● mainly **carbon dioxide** (like the atmospheres of Mars and Venus today) ● **nitrogen** (which gradually built up in the atmosphere) ● **water vapour** (which condenses to form the oceans) ● **sulfur dioxide** and small amounts of **methane** and **ammonia**
4 billion years ago	Decrease in carbon dioxide and other gases / Increase in oxygen and nitrogen	Green plants and algae evolve and… ● **carbon dioxide** is reduced as the plants take it in and give out **oxygen** by photosynthesis: $6CO_2 + 6H_2O \rightarrow C_6H_{12}O_6 + 6O_2$. The percentage of oxygen gradually increased to a level that enabled animals to evolve
3 billion years ago	Other gases / O_2 / CO_2 / N_2	● microorganisms that cannot tolerate oxygen are killed off ● **carbon** from carbon dioxide becomes locked up in **sedimentary rocks** formed from the shells and skeletons of marine organisms ● other gases react with oxygen to release **nitrogen** ● nitrogen is also produced by bacteria removing nitrates from decaying plant material
2 billion years ago	Decrease in carbon dioxide / Increase in nitrogen and oxygen	There is now about 20% **oxygen** and about 80% **nitrogen** in the atmosphere
1 billion years ago	Other gases / CO_2 / O_2 / N_2	The amount of **carbon dioxide** has decreased significantly.
Now		

Composition of the atmosphere today

The proportions of gases in the atmosphere have been more or less the same for about 200 million years. **Water vapour** may also be present in varying quantities (0–3%).

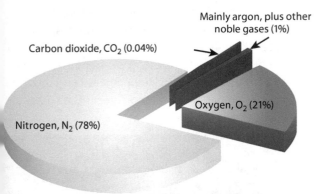

Mainly argon, plus other noble gases (1%)

Carbon dioxide, CO_2 (0.04%)

Oxygen, O_2 (21%)

Nitrogen, N_2 (78%)

How carbon dioxide decreased

The amount of carbon dioxide in the atmosphere today is much less than it was when the atmosphere first formed. This is because…

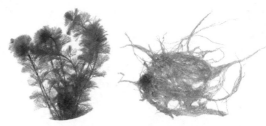

● green plants and algae use carbon dioxide for photosynthesis

● carbon dioxide is used to form sedimentary rocks, e.g. limestone

● fossil fuels such as oil and coal (a sedimentary rock made from thick plant deposits that were buried and compressed at high temperatures over millions of years) have captured CO_2.

SUMMARY

● Although theories differ, it is thought the Earth was formed 4.6 billion years ago and the atmosphere was mostly carbon dioxide.

● The composition of the atmosphere has changed over time and is now mostly nitrogen.

● Carbon dioxide decreased due to it being used in photosynthesis, forming sedimentary rocks and CO_2 being captured as fossil fuels.

QUESTIONS

QUICK TEST

1. Explain how oxygen became present in the atmosphere.

2. Name two gases that volcanoes released into the early atmosphere.

3. How much of the atmosphere today is water vapour?

4. What gas do green plants and algae use for photosynthesis?

EXAM PRACTICE

1. Describe and explain the changes that green plants and algae have had on the composition of the atmosphere since their evolution. **[3 marks]**

2. A sample of air was found to have the following composition.

   Total volume = 820 cm³

   Volume of oxygen = 164 cm³
   Volume of nitrogen = 640 cm³
   Volume of other gases =

   a) Calculate the volume of other gases in the sample of air. **[1 mark]**

   b) Name two gases that are present in the other gases. **[2 marks]**

   c) Calculate the percentage of oxygen in the sample of air. **[1 mark]**

Climate change

WS Human activity and global warming

Some human activities increase the amounts of greenhouse gases in the atmosphere including…

Combustion of fossil fuels releases carbon dioxide into the atmosphere

Increased animal farming releases more methane into the atmosphere, e.g. as a by-product of digestion and decomposition of waste

Deforestation reduces the amount of carbon dioxide removed from the atmosphere by photosynthesis

Decomposition of rubbish in landfill sites also releases methane into the atmosphere

Greenhouse gases

The temperature on Earth is maintained at a level to support life by the greenhouse gases in the atmosphere. These gases allow short wavelength radiation from the Sun to pass through but absorb the long wavelength radiation reflected back from the ground trapping heat and causing an increase in temperature. Common greenhouse gases are water vapour, carbon dioxide and methane.

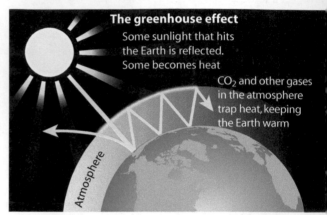

The greenhouse effect

Some sunlight that hits the Earth is reflected. Some becomes heat

CO_2 and other gases in the atmosphere trap heat, keeping the Earth warm

Atmosphere

The increase in carbon dioxide levels in the last century or so correlates with the increased use of fossil fuels by humans.

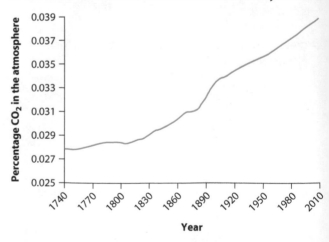

Based on peer-reviewed evidence, many scientists believe that increasing these human activities will lead to global climate change.

Predicting the impact of changes on global climate change is not easy because of the many different contributing factors involved. This can lead to simplified models and speculation often presented in the media that may not be based on all of the evidence. Some people misrepresent the evidence to suit their own needs; this can lead to bias.

Global climate change

Increasing average global temperature is a major cause of climate change. Potential effects of climate change include:

| Rising sea levels, which may cause flooding and coastal erosion | Changes in the food producing capacity of some regions | More frequent and severe storms | Changes to the amount, timing and distribution of rainfall | Changes to the distribution of wildlife species | Temperature and water stress for humans and wildlife |

WS Reducing the carbon footprint

The carbon footprint is a measure of the total amount of carbon dioxide (and other greenhouse gases) emitted over the life cycle of a product, service or event.

Problems of trying to reduce the carbon footprint include…

- disagreement over the causes and consequences of global climate change
- lack of public information and education
- lifestyle changes, e.g. greater use of cars / aeroplanes
- economic considerations, i.e. the financial costs of reducing the carbon footprint
- incomplete international co-operation.

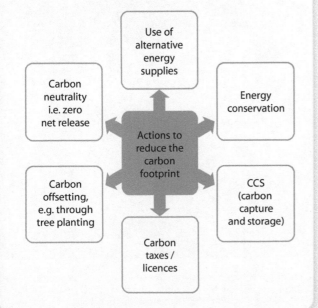

SUMMARY

- **Greenhouse gases maintain the temperature on Earth.**
- **Human activities increase the amount of greenhouse gases, leading to climate change and global warming.**

QUESTIONS

QUICK TEST

1. Name two greenhouse gases.
2. Suggest two potential effects of global climate change.
3. What is a carbon footprint?

EXAM PRACTICE

1. Explain how human activity can contribute to the emission of the greenhouse gases methane and carbon dioxide. **[3 marks]**

Atmospheric pollution

Pollutants from fuels

The combustion of fossil fuels is a major source of atmospheric pollutants. Most fuels contain carbon and often sulfur is present as an impurity. Many different gases are released into the atmosphere when a fuel is burned.

Solid particles and unburned hydrocarbons can also be released forming particulates in the air.

carbon monoxide

carbon dioxide

Gases produced by burning fuels

oxides of nitrogen

water vapour

sulfur dioxide

Sulfur dioxide is produced by the oxidation of sulfur present in fuels – often from coal-burning power stations

Carbon monoxide and soot (carbon) are produced by incomplete combustion of fuels

Oxides of nitrogen are formed from the reaction between nitrogen and oxygen from the air – often from the high temperatures and sparks in the engines of motor vehicles

Properties and effects of atmospheric pollutants

Carbon monoxide is a colourless, odourless toxic gas and so it is difficult to detect. It combines with haemoglobin in the blood, which reduces the oxygen-carrying capacity of blood.

Sulfur dioxide and **oxides of nitrogen** cause respiratory problems in humans and can form acid rain in the atmosphere. Acid rain damages plants and buildings.

Particulates in the atmosphere can cause global dimming, which reduces the amount of sunlight that reaches the Earth's surface. Breathing in particulates can also damage lungs, which can cause health problems.

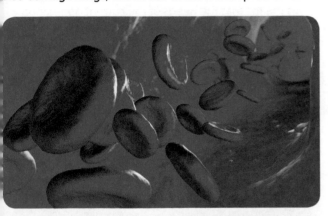

SUMMARY

- Combustion of fossil fuels is a major source of atmospheric pollutants.
- Gases produced by burning fuels include carbon monoxide, carbon dioxide, sulfur dioxide, water vapour and oxides of nitrogen.

QUESTIONS

QUICK TEST

1. Name three gases produced by burning fuels.

2. How is sulfur dioxide formed?

EXAM PRACTICE

1. Cars over three years old are required to have an annual test to check the roadworthiness of the vehicle.

   As part of this test the amounts of the following pollutants in exhaust gases are measured:

   Carbon monoxide
   Hydrocarbons
   Oxides of nitrogen
   Particulates

   a) Suggest why carbon monoxide may be present in exhaust gases. **[1 mark]**

   b) Suggest why older cars may have higher levels of hydrocarbons in their exhaust gases than new cars. **[1 mark]**

   c) Why is it harmful to breathe in oxides of nitrogen? **[1 mark]**

   d) Why are particulates in the atmosphere harmful? **[2 marks]**

Using the Earth's resources and obtaining potable water

Earth's resources

We use the Earth's **resources** to provide us with warmth, shelter, food and transport. These needs are met from natural resources which, supplemented by agriculture, provide food, timber, clothing and fuels. Resources from the earth, atmosphere and oceans are processed to provide energy and materials.

Example

Chemistry plays a role in providing sustainable development. This means that the needs of the current generation are met without compromising the potential of future generations to meet their own needs. For example…

Industrial chemical processes can be used to make new materials, reducing the demand for natural resources

Chemistry plays an important role in improving agricultural processes, e.g. by developing fertilisers

Drinking water

Water that is safe to drink is called potable water. It is not pure in the chemical sense because it contains dissolved minerals and ions.

Water of appropriate quality is essential for life. This means that it contains sufficiently low levels of dissolved salts and microbes.

In the UK, most potable water comes from rainwater. To turn rainwater into potable water, water companies carry out a number of processes.

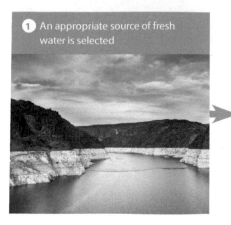

1 An appropriate source of fresh water is selected

2 It is then passed through filter beds to remove any solid impurities

3 Finally it is sterilised (suitable sterilising agents include chlorine, ozone and ultraviolet light) to kill microbes, making it safe to drink

Waste water

Urban lifestyles and industrial processes generate large quantities of waste water, which requires treatment before being released into the environment. Sewage and agricultural waste water require removal of organic matter and harmful microbes. Industrial waste water may require removal of organic matter and harmful chemicals.

Sewage treatment includes…

- screening and grit removal
- sedimentation to produce sewage sludge and effluent
- anaerobic digestion of sewage sludge
- aerobic biological treatment of effluent.

When supplies of fresh water are limited, removal of salt (desalination) of salty water / seawater can be used.

This is done in two ways:

- By distillation.

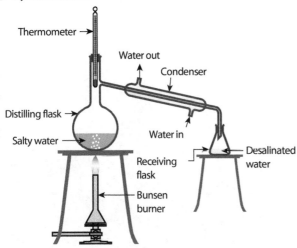

- By processes that use membranes, such as reverse osmosis.

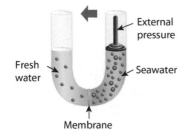

However, these processes require large amounts of energy.

SUMMARY

- **Humans use the Earth's resources to supply them with their needs.**
- **Water is an important resource. Water goes through processes to become safe to drink (potable water).**
- **Modern life creates a lot of waste water, which must be treated before being released into the environment.**

QUESTIONS

QUICK TEST

1. Explain the meaning of the term 'sustainable development'.

2. What is the difference between pure water and potable water?

3. Outline the main stages in the treatment of sewage.

EXAM PRACTICE

1. The diagram below shows how water can be separated from sea water.

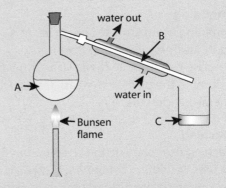

a) Name this method of desalination. **[1 mark]**

b) Identify the labels A, B and C. **[3 marks]**

c) Name another method of removing salt from sea water. **[1 mark]**

d) Suggest why both of these methods are expensive to carry out. **[1 mark]**

HT Alternative methods of extracting metals

Extracting copper

Copper is an important metal with lots of uses. It is used in electrical wiring because it is an excellent conductor of electricity. It is also used in water pipes because it conducts heat and does not corrode or react with the water.

copper ore

The Earth's resources of metal ores are limited and copper ores are becoming scarce.

New ways of extracting copper from low-grade ores include…

- **phytomining**
  Phytomining uses plants to absorb metal compounds. This means that the plants accumulate metal within them. Harvesting and then burning the plants leaves ash that is rich in the metal compounds.

- **bioleaching**
  Bioleaching uses bacteria to extract metals from low-grade ores. A solution containing bacteria is mixed with a low-grade ore. The bacteria release the metals into solution (known as a leachate) where they can be easily extracted.

These new extraction methods reduce the impact on the Earth of mining, moving and disposing of large amounts of rock.

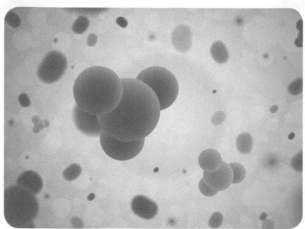

Processing metal compounds

The metal compounds from phytomining and bioleaching are processed to obtain the metal. Copper can be obtained from solutions of copper compounds by…

- **displacement** using scrap iron – iron is more reactive than copper, so placing iron into a solution of copper will result in copper metal being displaced.

Displacement

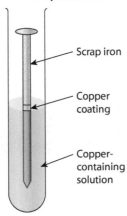

- Scrap iron
- Copper coating
- Copper-containing solution

- **electrolysis.**

Electrolysis

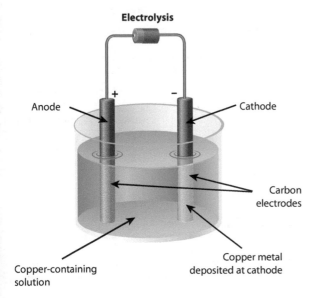

- Anode
- Cathode
- Carbon electrodes
- Copper-containing solution
- Copper metal deposited at cathode

SUMMARY

- **Copper is a metal that has many uses.**
- **Copper can be extracted from its ore by phytomining or bioleaching.**
- **Copper can be obtained from solutions of copper compounds by displacement or electrolysis.**

QUESTIONS

QUICK TEST

1. Suggest one use of copper.

2. Name two ways of extracting copper from low-grade ones.

3. What is bioleaching?

4. Give one way that metal can be extracted from a metal containing solution.

EXAM PRACTICE

1. **a)** Explain why alternative methods of extracting copper such as bioleaching are being developed. **[2 marks]**

   b) A solution containing copper ions is obtained from a bioleaching plant.

   Describe two ways that copper metal can be obtained from this solution. **[2 marks]**

   c) Other than mining and bioleaching, name another method of obtaining copper from ores. **[1 mark]**

Life-cycle assessment and recycling

ⓦⓢ Life-cycle assessments

Life-cycle assessments (LCAs) are carried out to evaluate the environmental impact of products in each of the following stages.

| Extracting and processing raw materials | Manufacturing and packaging | Disposal at the end of useful life | Transport and distribution at each of the previous stages |

The following steps are considered when carrying out an LCA.

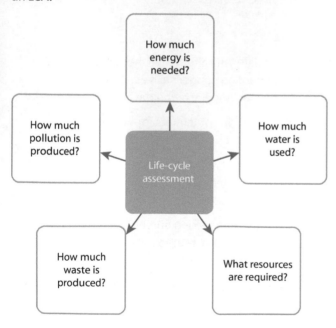

It is not always easy to obtain accurate figures. This means that selective or abbreviated LCAs, which are open to bias or misuse, can be devised to evaluate a product, to reinforce predetermined conclusions or to support claims for advertising purposes.

For example, look at the LCA below.

Example of an LCA for the use of plastic (polythene) and paper shopping bags		
	Amount per 1000 bags over the whole LCA	
	Paper	Plastic (polythene)
Energy use (MJ)	2590	713
Fossil fuel use (Kg)	28	13
Solid waste (Kg)	34	6
Greenhouse gas emissions (kg CO_2)	72	36
Freshwater use (litres)	3387	198

This LCA provides evidence supporting the argument that using polythene bags is better for the environment than paper bags!

Many of these values are relatively easy to quantify. However, some values, such as the amount of pollution, are often difficult to measure and so value judgements have to be made. This means that carrying out an LCA is not a purely objective process.

Ways of reducing the use of resources

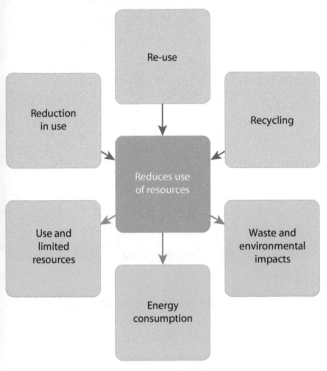

Many materials such as glass, metals, building materials, plastics and clay ceramics are produced from limited raw materials. Most of the energy used in their production comes from limited resources, such as fossil fuels. Obtaining raw materials from the earth by quarrying and mining has a detrimental environmental impact.

Some products, like glass, can be **reused** (e.g. washing and then using again for the same purpose). Recycled glass is crushed, melted and remade into glass products.

Other products cannot be reused and so they are **recycled** for a different use.

Metals are recycled by sorting them, followed by melting them and recasting / reforming them into different products. The amount of separation required for recycling depends on the material (e.g. whether they are magnetic or not) and the properties required of the final product. For example, some scrap steel can be added to iron from a blast furnace to reduce the amount of iron that needs to be extracted from iron ore.

SUMMARY

- LCAs are carried out to evaluate the environmental impact of products at each stage in their production and distribution.
- Recycling is important in order to reduce the use of resources, and therefore reduce energy use, waste and environmental impacts.

QUESTIONS

QUICK TEST

1. What does a life-cycle assessment measure?

2. State two factors that a life-cycle assessment tries to evaluate.

3. What is a blast furnace?

4. Suggest one way in which we can reduce the use of resources.

EXAM PRACTICE

1. A Life Cycle Assessment (LCA) comparing paper and plastic shopping bags was carried out by an independent scientist.

   The report stated:

   'Production of a paper bag produces 2.5 times the amount of carbon dioxide than a plastic bag.'

   A journalist quoted this fact in an article and wrote:

   'Based on this fact, on environmental grounds we should be using plastic bags.'

   Evaluate this statement using the information provided and your own knowledge.   **[3 marks]**

Forces

Forces

Scalar quantities have magnitude only.

Vector quantities have magnitude and an associated direction.

A vector quantity can be represented by an arrow. The length of the arrow represents the magnitude, and the direction of the arrow represents the direction of the vector quantity.

Contact and non-contact forces

A force is a push or pull that acts on an object due to the interaction with another object. Force is a vector quantity. All forces between objects are either:

- contact forces – the objects are physically touching, for example: friction, air resistance, tension and normal contact force

  or

- non-contact forces – the objects are physically separated, for example: gravitational force, electrostatic force and magnetic force.

Air resistance

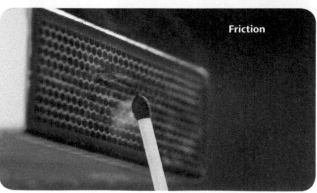

Friction

Gravity

- Weight is the force acting on an object due to gravity.
- All matter has a gravitational field that causes attraction. The field strength is much greater for massive objects.
- The force of gravity close to the Earth is due to the gravitational field around the Earth.
- The weight of an object depends on the gravitational field strength at the point where the object is.
- The weight of an object and the mass of an object are directly proportional (weight ∝ mass). Weight is a vector quantity as it has a magnitude and a direction. Mass is a scalar quantity as it only has a magnitude.

Weight is measured using a calibrated spring-balance – a **newtonmeter**.

Weight can be calculated using the following equation:

> **weight = mass × gravitational field strength**
>
> $$W = mg$$
>
> - weight, W, in newtons, N
> - mass, m, in kilograms, kg
> - gravitational field strength, g, in newtons per kilogram, N/kg

> **Example**
>
> What is the weight of an object with a mass of 54 kg in a gravitational field strength of 10 N/kg?
>
> weight = mass × gravitational field strength
>
> = 54 kg × 10 N/kg
>
> = 540 Nm

Resultant forces

The resultant force equals the total effect of all the different forces acting on an object.

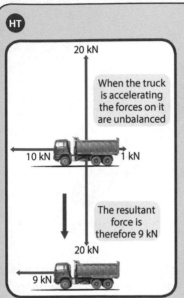

When the truck is accelerating the forces on it are unbalanced

If the forces on an object are balanced, the resultant force is zero. If the object is stationary it remains stationary and if it is moving it continues moving at a constant speed.

The resultant force is therefore 9 kN

A single force can be resolved into two components acting at right angles to each other. The two component forces together have the same effect as the single force.

Work done and energy transfer

Work is done when a force causes an object to move. The force causes a displacement.

The work done by a force on an object can be calculated using the following equation:

work done = force × **distance moved along the line of action of the force**

$$W = Fs$$

- work done, W, in joules, J
- force, F, in newtons, N
- distance, s, in metres, m (s represents displacement, commonly called distance)

Example

What work is done when a force of 90 N moves an object 14 m?

work done = force × distance moved along the line of action of the force

work done = 90 × 14 = 1260 J

One joule of work is done when a force of one newton causes a displacement of one metre.

1 joule = 1 newton metre

Work done against the frictional forces acting on an object causes a rise in the temperature of the object.

SUMMARY

- Force is a vector quantity. A force is a push or pull that acts on an object due to the interaction with another object.
- Resultant force equals the total effect of all the different forces acting on an object.
- Work is done when a force causes an object to move.

QUESTIONS

QUICK TEST

1. What is a vector quantity?

2. What is a scalar quantity?

3. What is the weight of a 33 g object in a gravitational field strength of 10 N/kg?

4. What is 57 Nm in joules?

EXAM PRACTICE

1. In a warehouse a metal crate is pushed 4m along the floor.

   a) The weight of the crate is 560N.

      Assuming a gravitational field strength of 9.8 N/kg what is the mass of the crate? **[2 marks]**

   b) What work is done pushing this crate? **[2 marks]**

   c) Pushing the crate is an example of exerting a contact force; if the crate was pulled by a magnet this would be an example of a non-contact force. Explain this difference. **[2 marks]**

Forces and elasticity

Plastic Deformation

When an object is stretched and returns to its original length after the force is removed, it is **elastically deformed**.

When an object does not return to its original length after the force has been removed, it is **inelastically deformed**. This is **plastic deformation**.

Extension

The extension of an elastic object, such as a spring, is directly proportional to the force applied (extension ∝ force applied), if the limit of proportionality is not exceeded.

- **Stretching** – when a spring is stretched, the force pulling it exceeds the force of the spring.
- **Bending** – when a shelf bends under the weight of too many books, the downward force is from the weight of the books, and the shelf is resisting this weight.
- **Compressing** – when a car goes over a bump in the road, the force upwards is opposed by the springs in the car's suspension.

Spring balance

In order for any of these processes to occur, there must be more than one force applied to the object to bring about the change.

The force on a spring can be calculated using the following equation:

force = spring constant × extension

$$F = ke$$

- force, F, in newtons, N
- spring constant, k, in newtons per metre, N/m
- extension, e, in metres, m

Example

A spring with spring constant 35 N/m is extended by 0.3 m. What is the force on the spring?

$F = ke$

$\quad = 35 \times 0.3$

$\quad = 10.5$ N

This relationship also applies to the compression of an elastic object, where the extension, e, would be the compression of the object.

A force that stretches (or compresses) a spring does **work** and **elastic potential energy** is stored in the spring. Provided the spring does not go past the limit of proportionality, the work done on the spring equals the stored elastic potential energy.

Force and extension

Force and extension have a linear relationship.

If force and extension are plotted on a graph, the points can be connected with a straight line (see Graph 1).

The points in a non-linear relationship cannot be connected by a straight line (see Graph 2).

Graph 1 – Linear relationship

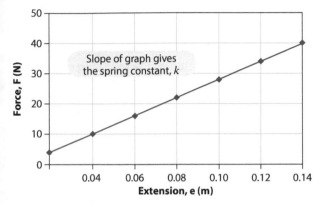

Graph 2 – Non-linear relationship

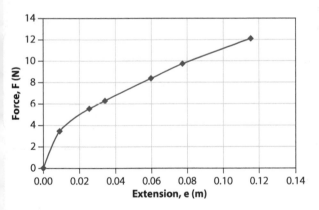

In order to calculate the spring constant in Graph 1, take two points and apply the equation.

So, spring constant $= \frac{\text{force}}{\text{extension}}$

$= \frac{10}{0.04}$

$= 250$ N/m

SUMMARY

● **An object that returns to its original length when the stretching force is removed is called elastically deformed.**

● **An object that doesn't return to its original length when the stretching force is removed is called inelastically deformed.**

● **Force and extension have a linear relationship if the limit of proportionality has not been reached.**

QUESTIONS

QUICK TEST

1. What is meant by elastically deformed?

2. What is the force of a spring with an extension of 0.3 m and a spring constant of 2 N/m?

3. What type of energy is stored in a spring?

EXAM PRACTICE

1. An investigation was carried out into the deformation of an elastic band.

   a) The elastic band is stretched by a force of 7N. The spring constant of the elastic band is 18 N/m.

   How far is the elastic band extended? **[2 marks]**

   b) The elastic band was stretched again with a number of different forces. Force and extension were then plotted on a graph. The band did not exceed the limit of proportionality.

   i) Predict what type of line could be used to connect the points on the graph. **[1 mark]**

   ii) What type of relationship is shown by these results? **[1 mark]**

   iii) If the band had exceeded the limits of proportionality, would the results show a different relationship?

   Explain your answer. **[2 marks]**

Speed and velocity

Distance and speed

Distance is how far an object moves. As distance does not involve direction, it is a scalar quantity.

Displacement includes both the distance an object moves, measured in a straight line from the start point to the finish point, and the direction of that straight line. As displacement has magnitude and a direction, it is a vector quantity.

The speed of a moving object is rarely constant. When people walk, run or travel in a car, their speed is constantly changing.

The speed that a person can walk, run or cycle depends on:

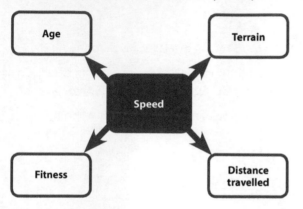

Some typical speeds are:

- walking – 1.5 m/s
- running – 3 m/s
- cycling – 6 m/s
- car – 20 m/s
- train – 35 m/s

As speed has a magnitude but not a direction, it is a scalar quantity. The speed of sound and the speed of wind also vary.

A typical value for the speed of sound in air is 330 m/s. A gale force wind is one with a speed above 14 m/s.

Speed can be investigated in labs using equipment such as light gates.

For an object travelling at a constant speed, distance travelled can be calculated by the following equation:

distance travelled = speed × time

$$s = vt$$

- distance, s, in metres, m
- speed, v, in metres per second, m/s
- time, t, in seconds, s

Example

What distance is covered by a runner with a speed of 3 m/s in 6000 seconds?

distance travelled $= $ speed $\times$ time

$= 3 \times 6000$

$= 18\,000$ m

$= 1.8$ km

Velocity

The velocity of an object is its speed in a given direction. As velocity has a magnitude and a direction, it is a vector quantity.

HT A car travelling at a constant speed around a bend would have a varying velocity because it is changing direction.

When an object moves in a circle, the direction of the object is continually changing. This means that an object moving in a circle at constant speed, such as an orbiting satellite, has a continually changing velocity. For motion in a circle, there is a resultant centripetal force that acts towards the centre of the circle.

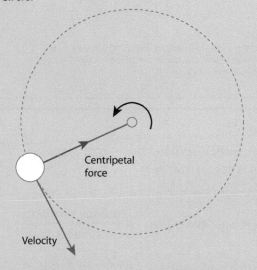

Centripetal force

Velocity

SUMMARY

- **Distance is how far an object moves.**
- **Displacement is how far an object moves, and the direction it moves in.**
- **Velocity is the speed of an object in a given direction.**
- **Velocity is a vector quantity as it has a magnitude and direction.**

QUESTIONS

QUICK TEST

1. Explain why speed is a scalar quantity.

2. What is the speed of a walker who covers 15 km in 3 hours?

3. What is the typical speed of sound in air?

EXAM PRACTICE

1. **a)** A motorbike and rider travel at a speed of 21m/s for a time of 56 seconds.

    How far do the motorbike and rider travel in this time? **[2 marks]**

    b) The rider decelerates to a speed of 15m/s for a distance of 14 km.

    How long does it take to travel this distance?

    Give your answer in minutes. **[3 marks]**

    HT c) The rider maintains this speed as she goes round a corner.

    What happens to her velocity as she goes around the corner?

    Explain your answer. **[2 marks]**

Distance–time and velocity–time graphs

Distance and time

The distance an object moves in a straight line can be represented by a distance–time graph.

The speed of an object can be calculated from the gradient of its distance–time graph.

The graph below shows a person jogging.

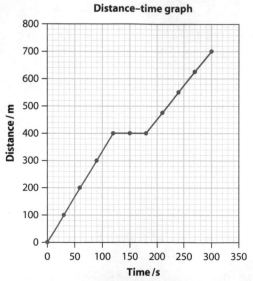

Distance–time graph

The average speed of this jogger between 0–120 seconds can be worked out as follows:

$$\text{speed} = \frac{\text{distance}}{\text{time}} = \frac{400}{120} = 3.33 \text{ m/s}$$

> **HT** The speed of an accelerating object can be determined by using a tangent to measure the gradient of the distance–time graph.

Acceleration

Acceleration can be calculated using the following equation:

> **average acceleration** = $\dfrac{\text{change in velocity}}{\text{time taken}}$
>
> $$\left[a = \frac{\Delta v}{t} \right]$$

- acceleration, a, in metres per second squared, m/s^2
- change in velocity, Δv, in metres per second, m/s
- time, t, in seconds, s

Example

A car travels from 0 to 27 m/s in 3.8 seconds. What is its acceleration?

change in velocity $= 27 - 0 = 27$ m/s

average acceleration $= \dfrac{\text{change in velocity}}{\text{time taken}}$

$= \dfrac{27}{3.8} = 7.1$ m/s^2

An object that slows down (decelerates) has a negative acceleration.

Velocity–time graph

Acceleration can be calculated from the gradient of a velocity–time graph.

> **HT** Distance travelled can be calculated from the area under a velocity–time graph.

The graph at the top of the following page shows the movement of a car.

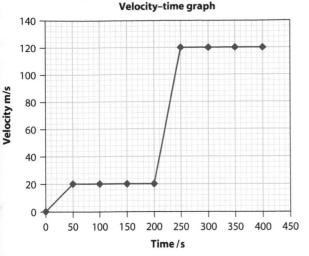

Velocity–time graph

The acceleration between 0–50 seconds can be worked out as follows:

$$\text{gradient of line} = \frac{\text{velocity}}{\text{time}}$$

$$= \frac{20}{50} = 0.4 \text{ m/s}^2$$

As the gradient of the line is steeper between 200–250 seconds than it is between 0–50 seconds, the acceleration must have been greater between 200–250 seconds.

HT The distance travelled between 0–200 seconds can be worked out as follows:

Area from 0–50s

$$= (20 \times \frac{50}{2}) = 500 \text{ m}$$

Area from 50–200s

$$= 20 \times 150 = 3000 \text{ m}$$

Total distance travelled $= 3500$ m

The equation below applies to uniform motion:

$$\left(\begin{array}{c} \textbf{final} \\ \textbf{velocity}^2 \end{array} - \begin{array}{c} \textbf{initial} \\ \textbf{velocity}^2 \end{array} \right) = 2 \times \textbf{acceleration} \times \textbf{distance}$$

$$v^2 - u^2 = 2as$$

- final velocity, v, in metres per second, m/s
- initial velocity, u, in metres per second, m/s
- acceleration, a, in metres per second squared, m/s^2
- distance, s, in metres, m

Falling objects

Near the Earth's surface, any object falling freely under gravity has an acceleration of about 10 m/s^2 (9.8m/s^2). This is the gravitational field strength, or acceleration due to gravity, and is used to calculate weight.

An object falling through a fluid initially accelerates due to the force of gravity. Eventually the resultant force will be zero and the object will move at its terminal velocity.

SUMMARY

- **Distance–time graphs show the movement of an object in a straight line.**
- **Velocity–time graphs can be used to calculate acceleration of an object.**
- **Falling objects accelerate due to the force of gravity until they reach terminal velocity.**

QUESTIONS

QUICK TEST

1. What is acceleration?

2. What type of acceleration will an object have when it is slowing down?

3. What does the gradient of a distance–time graph show?

EXAM PRACTICE

1. A skydiver's initial velocity is zero. When he jumps from the plane he accelerates to 54m/s in 16 seconds. After this point he doesn't accelerate any further.

   a) Calculate the skydiver's average acceleration. **[2 marks]**

   b) Explain why the skydiver doesn't accelerate further. **[2 marks]**

Newton's laws

Newton's first law

Newton's first law deals with the effect of resultant forces.

- If the resultant force acting on an object is zero and the object is stationary, the object remains stationary.

- If the resultant force acting on an object is zero and the object is moving, the object continues to move at the same speed and in the same direction (its velocity will stay the same).

Newton's first law means that the velocity of an object will only change if a resultant force is acting on the object.

Examples of resultant force:

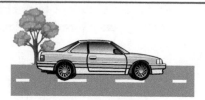

A car is stationary. At this point the resultant force is zero.

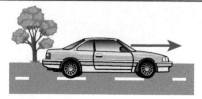

The car starts to move and accelerates. The car is accelerating as the resultant force on it is no longer zero.

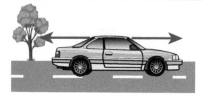

The car travels at a constant velocity. The resultant force is zero again.

> **HT** The tendency of objects to continue in their state of rest or of uniform motion is called inertia.

Stationary object – zero resultant force

Object moving at constant speed – zero resultant force

Newton's second law

The acceleration of an object is proportional to the resultant force acting on the object, and inversely proportional to the mass of the object.

Therefore:

> **acceleration ∝ resultant force**
>
> **resultant force = mass × acceleration**
> - force, F, in newtons, N
> - mass, m, in kilograms, kg
> - acceleration, a, in metres per second squared, m/s^2

> **Example**
>
> A motorbike and rider of mass 270 kg accelerate at 6.7 m/s^2. What is the resultant force on the motorbike?
>
> $$270 \times 6.7 = 1809 \text{ N}$$

If the motorbike slows to a constant speed, the resultant force would now be 0 (as acceleration = 0, 270 × 0 = 0).

> **Inertial mass** is a measure of how difficult it is to change the velocity of an object. It is defined by the ratio of force over acceleration.

Newton's third law

Whenever two objects interact, the forces they exert on each other are equal and opposite.

When a fish swims it exerts a force on the water, pushing it backwards. The water exerts an equal and opposite force on the fish, pushing it forwards.

SUMMARY

- Newton's first law says that the velocity of an object will only change if a resultant force is acting on the object.
- Newton's second law says that the acceleration of an object is proportional to the resultant force acting on it, and inversely proportional to its mass.
- Newton's third law says that when two objects interact, the forces they exert on each other are equal and opposite.

QUESTIONS

QUICK TEST

1. A sprinter of mass 93 kg accelerates at 10 m/s^2. What is the resultant force?

2. A book of weight 83 N is on a desk. What force is the desk exerting on the book?

3. What happens to the speed of a moving object that has a resultant force of 0 acting on it?

4. What happens to the speed of a stationary object that has a resultant force of 0 acting on it?

EXAM PRACTICE

1. A cannonball with a mass of 7kg when fired from a cannon experiences a force of 2100N.

   a) What is the acceleration of the cannonball? **[2 marks]**

   b) What force does the cannon experience?

      Explain how you arrived at your answer. **[3 marks]**

Forces, braking and momentum

Stopping distance

The stopping distance of a vehicle is a sum of the thinking distance and the braking distance.

> **stopping distance =** **thinking distance (the reaction time of the driver)** + **braking distance (distance the vehicle travels under the braking force)**

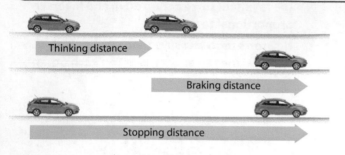

Thinking distance

Braking distance

Stopping distance

A greater vehicle speed leads to a greater stopping distance (given a set braking force). As the vehicle is going faster, the car will travel further in the time taken for the driver to react and apply the brakes.

The graph below shows the stopping distances over a range of speeds for a car.

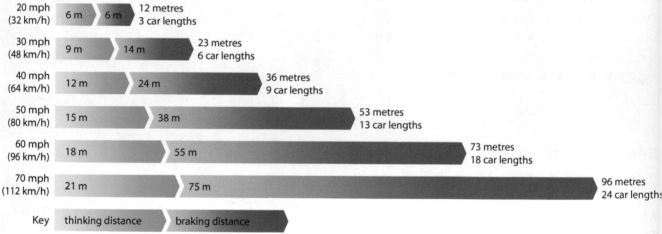

Typical stopping distances

Speed	Thinking distance	Braking distance	Total
20 mph (32 km/h)	6 m	6 m	12 metres — 3 car lengths
30 mph (48 km/h)	9 m	14 m	23 metres — 6 car lengths
40 mph (64 km/h)	12 m	24 m	36 metres — 9 car lengths
50 mph (80 km/h)	15 m	38 m	53 metres — 13 car lengths
60 mph (96 km/h)	18 m	55 m	73 metres — 18 car lengths
70 mph (112 km/h)	21 m	75 m	96 metres — 24 car lengths

Key: thinking distance / braking distance

Braking distance

The braking distance of a vehicle can be affected by the vehicle's mass, adverse conditions outside and poor vehicle condition.

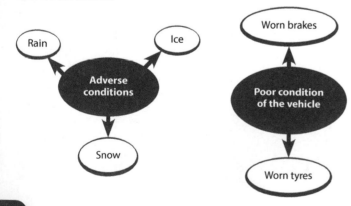

Rain — Ice — **Adverse conditions** — Snow

Worn brakes — **Poor condition of the vehicle** — Worn tyres

Reaction time

The average reaction time is 0.2–0.9 seconds but this varies from person to person. Thinking distance increases as speed increases. This is because a faster vehicle travels further during the time taken to react.

A driver's reaction time can be affected by tiredness, drugs and alcohol. Distractions, such as eating, drinking or using a mobile phone, may also affect a driver's ability to react.

Braking force

To slow a vehicle down, the friction between the tyres and the road must be increased. This is done by the brakes.

The following flow chart shows what happens when a force is applied to the brakes of a vehicle.

Work is done by the friction force between the brakes and the wheel.	→	This reduces the kinetic energy of the vehicle.	→	The temperature of the brakes increases.

- The greater the speed of a vehicle, the greater the braking force needed to stop the vehicle in a certain distance.
- The greater the braking force, the greater the deceleration of the vehicle.
- The work done to stop the vehicle is equal to the initial kinetic energy of the vehicle.

Large decelerations may lead to brakes overheating and/or loss of control.

HT Momentum

Moving objects have momentum. Momentum is given by the following equation:

momentum = mass × velocity

$$p = mv$$

- momentum, p, in kilograms metre per second, kg m/s
- mass, m, in kilograms, kg
- velocity, v, in metres per second, m/s

Example

A car of mass 1100 kg is travelling at 13 m/s. What is its momentum?

$p = mv$

$= 1100 \times 13$

$= 14\,300$ kg m/s

Conservation of momentum

In a closed system, the total momentum before an event is equal to the total momentum after the event. This is called **conservation of momentum**.

If two objects collide, their total momentum before the collision will equal their total momentum after the collision.

SUMMARY

- Stopping distance is the thinking distance and braking distance together.
- Average reaction time is 0.2–0.9 seconds, and can be affected by tiredness, drugs or alcohol.
- HT Moving objects have momentum. In a closed system, the total momentum before an event is equal to the total momentum after an event.

QUESTIONS

QUICK TEST

1. What two distances make up the stopping distance?

2. Give two examples of adverse road conditions which can increase the stopping distance.

3. Why is a large deceleration potentially dangerous?

EXAM PRACTICE

HT 1. A car is travelling at 7.5m/s. It has a mass of 1080kg.

   a) What is the momentum of the car? **[2 marks]**

   b) The car stops, several passengers get into the car and then it sets off, again travelling at 7.5 m/s with a momentum of 9675 kg m/s.

   What is the new mass of the car? **[2 marks]**

   c) This car is still travelling at 7.5 m/s when it is involved in a collision with a car which is stationary.

   What is the total momentum of both cars after the collision?

   Explain how you arrived at your answer. **[3 marks]**

Changes in energy

A system is an object or group of objects. When a system changes, so does the way energy is stored within its objects.

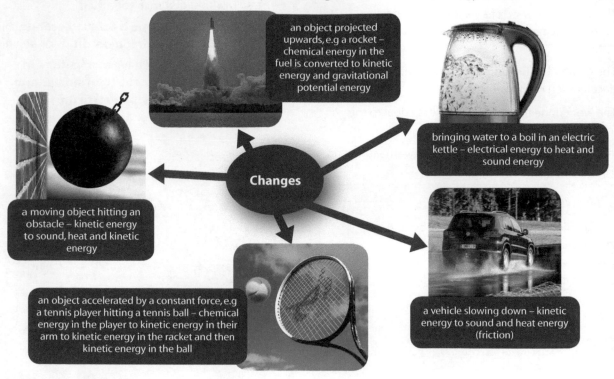

an object projected upwards, e.g a rocket – chemical energy in the fuel is converted to kinetic energy and gravitational potential energy

bringing water to a boil in an electric kettle – electrical energy to heat and sound energy

Changes

a moving object hitting an obstacle – kinetic energy to sound, heat and kinetic energy

an object accelerated by a constant force, e.g a tennis player hitting a tennis ball – chemical energy in the player to kinetic energy in their arm to kinetic energy in the racket and then kinetic energy in the ball

a vehicle slowing down – kinetic energy to sound and heat energy (friction)

Energy in moving objects

The kinetic energy of a moving object can be calculated using the following equation:

kinetic energy = 0.5 × mass × (speed)²

$$E_k = \frac{1}{2}mv^2$$

- kinetic energy, E_k, in joules, J
- mass, m, in kilograms, kg
- speed, v, in metres per second, m/s

Example

A ball of mass 0.3 kg falls at 10 m/s. What is the ball's kinetic energy?

$E_k = \frac{1}{2}mv^2$

$\quad = 0.5 \times 0.3 \times 10^2$

$\quad = 15$ J

Elastic potential energy

elastic potential energy = 0.5 × spring constant × extension²

$$E_e = \frac{1}{2}ke^2$$

(assuming the limit of proportionality has not been exceeded)

- elastic potential energy, E_e, in joules, J
- spring constant, k, in newtons per metre, N/m
- extension, e, in metres, m

Example

A spring has a spring constant of 500 N/m and is extended by 0.2 m. What is the elastic potential energy in the spring?

$E_e = \frac{1}{2}ke^2$

$\quad = \frac{1}{2}500 \times 0.2^2$

$\quad = 250 \times 0.04$

$\quad = 10$ J

Gravitational potential energy (g.p.e.)

The amount of gravitational potential energy gained by an object raised above ground level can be calculated using the following equation:

$$\text{g.p.e.} = \text{mass} \times \frac{\text{gravitational field strength}}{} \times \text{height}$$

$$E_p = mgh$$

- gravitational potential energy, E_p, in joules, J
- mass, m, in kilograms, kg
- gravitational field strength, g, in newtons per kilogram, N/kg
- height, h, in metres, m

Example

A car of mass 1500 kg drives up a hill which is 67 m high. What is the gravitational potential energy of the car at the top of the hill? (Assume the gravitational field strength is 10 N/kg)

$E_p = mgh$

$ = 15\,000 \times 67 \times 10$

$ = 10\,050$ kJ

Changes in thermal energy

The specific heat capacity of a substance is the amount of energy required to raise the temperature of one kilogram of the substance by one degree Celsius. The specific heat capacity can be used to calculate the amount of energy stored in or released from a system as its temperature changes. This can be calculated using the following equation:

$$\begin{array}{c}\text{change} \\ \text{in thermal} \\ \text{energy}\end{array} = \text{mass} \times \begin{array}{c}\text{specific} \\ \text{heat} \\ \text{capacity}\end{array} \times \begin{array}{c}\text{temperature} \\ \text{change}\end{array}$$

$$\Delta E = m\,c\,\Delta\Theta$$

- change in thermal energy, ΔE, in joules, J
- mass, m, in kilograms, kg
- specific heat capacity, c, in joules per kilogram per degree Celsius, J/kg °C
- temperature change, $\Delta\Theta$, in degrees Celsius, °C

Example

0.75 kg of 100°C water cools to 23°C. What is the change in thermal energy?

(Specific heat capacity of water = 4184 J/kg °C)

$\Delta E = mc\,\Delta\Theta$

$ = 0.75 \times 4184 \times (100 - 23)$

$ = 241\,626$ J $= 242$ kJ

SUMMARY

- Kinetic energy can be calculated using the equation: ½ x mass x (speed)²
- Gravitational potential energy can be calculated using the equation: mass x gravitational field strength x height
- Elastic potential energy can be calculated using the equation: 0.5 x spring constant x extension²
- The specific heat capacity of a substance is the amount of energy needed to raise the temperature of 1 kg by 1°C.

QUESTIONS

QUICK TEST

1. What is specific heat capacity?

2. What is the kinetic energy of a 620 g object travelling at 5 m/s?

3. What is the gravitational potential energy of a 17 kg object at a height of 456 m? (gravitational field strength = 10 N/kg). Give your answer to 4 significant figures.

EXAM PRACTICE

1. A slingshot is used to fire a ball diagonally upwards. The slingshot has a spring constant of 115 N/m and it is extended by 0.49m.

   a) Calculate the elastic potential energy of the slingshot. Give your answer to 3 significant figures. **[2 marks]**

   b) What energy transfer takes place when the slingshot is fired? **[2 marks]**

   c) The ball lands on the ground. What is its gravitational potential energy at this point?

   Explain your answer. **[2 marks]**

Conservation and dissipation of energy

Energy transfers in a system

Energy can be **transferred**, **stored** or **dissipated**. It cannot be created or destroyed.

In a closed system there is no net change to the total energy when energy is transferred.

Only part of the energy is usefully transferred. The rest of the energy dissipates and is transferred in less useful ways, often as heat or sound energy. Energy is then described as **wasted**.

Examples of processes which cause a rise in temperature and so waste energy as heat include:

● friction between the moving parts of a machine

● electrical work against the resistance of connecting wires.

If the energy that is wasted can be reduced, that means more energy can be usefully transferred. The less energy wasted, the more efficient the transfer.

Efficiency

The energy efficiency for any energy transfer can be calculated using the following equation:

$$\text{efficiency} = \frac{\text{useful output energy transfer}}{\text{total input energy transfer}}$$

Efficiency may also be calculated using the following equation:

$$\text{efficiency} = \frac{\text{useful power}}{\text{total power output}}$$

Example

Kat uses a hairdryer. Some of the energy is wasted as sound. The electrical energy input is 24 kJ. The energy wasted is 7 kJ. What is the efficiency of the hairdryer?

First calculate the useful energy transferred.

useful energy = total energy – wasted energy

= 24 – 7 = 17 kJ

Now calculate the efficiency.

$$\text{efficiency} = \frac{\text{useful output energy transfer}}{\text{total input energy transfer}}$$

$$= \frac{17}{24} = 0.71$$

This decimal efficiency can be represented as a percentage by multiplying it by 100.

$$0.71 \times 100 = 71\%$$

HT There are many ways energy efficiency can be increased:

● Lubrication, thermal insulation and low resistance wires reduce energy waste and improve efficiency.

● Thermal insulation, such as loft insulation, reduces heat loss.

● Low resistance wires reduce energy lost as heat when an electrical current flows through them.

Power

One of the definitions of power is work done over time (the rate at which work is done). The power equation is:

$$\text{power} = \frac{\text{work done}}{\text{time}} \qquad P = \frac{W}{T}$$

● P = power (watts)
● E = work done (joules)
● T = time (seconds)

Example

A weightlifter is lifting weights that have a mass of 80 kg. What power is required to lift them 2 metres vertically in 4 seconds? (Assume gravitational field strength of 10 N/kg)

$$\text{power} = \frac{\text{work done}}{\text{time}}$$

$$\text{work done} = \text{force} \times \text{distance}$$

$$\text{force} = \text{mass} \times \text{gravitational field strength}$$

$$= 80 \times 10 = 800 \text{ N}$$

$$\text{work done} = 800 \times 2 = 1600 \text{ J}$$

$$\text{power} = \frac{1600}{4}$$

$$= 400 \text{ W}$$

A second weightlifter lifted the same mass to the same height in 3 seconds. As he carried out the same amount of work but in a shorter time, he would have a greater power.

Thermal insulation

Thermal insulation has a low thermal conductivity, so has a slow rate of energy transfer by conduction. U-values give a measure of the heat loss through a substance. A higher U-value indicates that a material has a higher thermal conductivity.

Changing the material of walls to materials that have a lower thermal conductivity reduces heat loss (the U-value) and so a building cools more slowly, reducing heating costs. The graph below shows the heat lost by different types of wall.

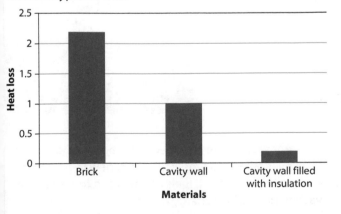

The house below is not fitted with insulation, so lots of heat is being lost through the walls due to their high thermal conductivity. This can be seen by the red/orange colour on the infrared image.

SUMMARY

- Energy can be transferred, stored or dissipated.
- Energy cannot be created or destroyed.
- Power means work done over time.
- Thermal insulation has a slow rate of energy transfer by conduction.

QUESTIONS

QUICK TEST

1. What is thermal insulation?

2. What is the power of a model crane which does 50 J of work in 2 seconds?

3. What effect does fitting cavity wall insulation have on the thermal conductivity of the building's wall?

EXAM PRACTICE

1. A washing machine transfers 600 J of useful energy out of a total of 802 J.

   What is the efficiency of the washing machine?   **[1 mark]**

National and global energy resources

Energy resources

Energy resources are used for transport, electricity generation and heating.

Energy resources can be divided into renewable and non-renewable. Renewable resources can be replenished as they are used. Non-renewable energy resources will eventually run out.

Renewable	Non-renewable
● Biofuel	● Coal
● Wind	● Oil
● Hydro-electricity	● Gas
● Geothermal	● Nuclear fuel
● The tides (tidal power)	
● The sun (solar power)	
● Water waves	

Renewable

Reliability

Fossil fuels (coal, oil and gas) and nuclear fuel are very reliable as they can always be used to release energy. Fossil fuels are burnt to release the stored chemical energy, and nuclear fuel radiates energy.

The wind and the Sun are examples of energy resources that are not very reliable, as the wind doesn't always blow and it's not always sunny.

Ethical and environmental concerns

● Burning fossil fuels produces carbon dioxide, which is a greenhouse gas. Increased greenhouse gas emissions are leading to climate change. Particulates and other pollutants are also released, which cause respiratory problems.

● Nuclear power produces hazardous nuclear waste and also has the potential for nuclear accidents, with devastating health and environmental consequences.

● Some people consider wind turbines to be ugly and to spoil the landscape.

● Building tidal power stations can lead to the destruction of important tidal habitats.

Non-renewable

In recent years, there have been attempts to move away from over-reliance on fossil fuels to renewable forms of energy. However, fossil fuels are still used to deliver the vast majority of our energy needs.

All the energy resources in the table on page 154 can be used to generate electricity. The following energy resources can be used for transport and heating.

Coal

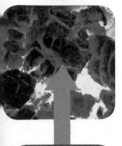

Oil

Heating

Natural Gas

Coal – steam trains

Wind – ships with sails

Transport

Solar – experimental solar-powered cars and planes are currently being tested

Oil – can be used to make petrol, diesel and jet fuel

Nuclear – submarines

SUMMARY

- **Energy resources can be renewable or non-renewable.**
- **Renewable energy resources can be replenished as they are used; non-renewable energy resources will eventually run out.**

QUESTIONS

QUICK TEST

1. Give three examples of renewable energy resources.

2. Give three examples of non-renewable energy sources.

3. Explain why solar and wind power are not reliable energy resources.

4. Describe an environmental concern about burning fossil fuels.

QUESTIONS

EXAM PRACTICE

1. **a)** Some people believe that a mix of renewable energy resources such as wind turbines, plus nuclear power plants, are the best solution to meet the UK's electricity generating needs.

   Explain the advantages of this. **[4 marks]**

   b) Why might some people protest against using nuclear power to meet the UK's energy needs? **[2 marks]**

   c) Some people might also protest against the building of wind farms and tidal power stations.

   Explain why. **[2 marks]**

Transverse and longitudinal waves

Transverse waves

In a transverse wave, the oscillations are perpendicular to the direction of energy transfer, such as the ripples on the surface of water.

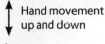

Hand movement up and down

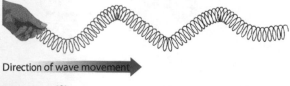

Direction of wave movement

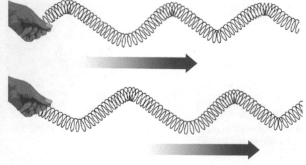

Longitudinal waves

In a longitudinal wave, the oscillations are parallel to the direction of energy transfer. Longitudinal waves show areas of compression and rarefaction, such as sound waves travelling through air.

Hand movement in and out

Compression

Expansion (rarefaction)

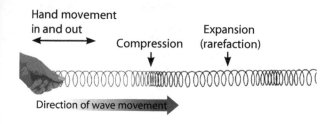

Direction of wave movement

Wave movement

In both sound waves in air and ripples on the water surface, it is the wave that moves forward rather than the air or water molecules. The waves transfer energy and information without transferring matter.

This can be shown experimentally.

For example, when a tuning fork is used to create a sound wave that moves out from the fork, the air particles don't move away from the fork. (This would create a vacuum around the tuning fork.)

Properties of waves

Waves are described by their:

- **amplitude** – The amplitude of a wave is the maximum displacement of a point on a wave away from its undisturbed position.

- **wavelength** – The wavelength of a wave is the distance from a point on one wave to the equivalent point on the adjacent wave.

- **frequency** – The frequency of a wave is the number of waves passing a point each second.

- **period** – The time for one complete wave to pass a fixed point. The equation for the time period (T) of a wave is given by the following equation:

$$\text{period} = \frac{1}{\text{frequency}}$$
$$\text{period} = \frac{1}{f}$$

- period, T, in seconds, s

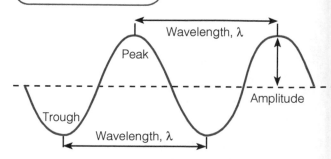

Wave speed

The wave speed or wave velocity is the speed at which the energy is transferred (or the wave moves) through the medium.

The wave speed is given by the following wave equation:

wave speed = frequency × wavelength
$$v = f \lambda$$

- wave speed, v, in metres per second, m/s
- frequency, f, in hertz, Hz
- wavelength, λ, in metres, m

Example

A sound wave in air has a frequency of 250 Hz and a wavelength of 1.32 m. What is the speed of the sound wave?

wave speed = frequency × wavelength

$\quad\quad\quad\quad = 250 \times 1.32$

$\quad\quad\quad\quad = 330$ m/s

A ripple tank and a stroboscope can be used to measure the speed of ripples on the surface of water.

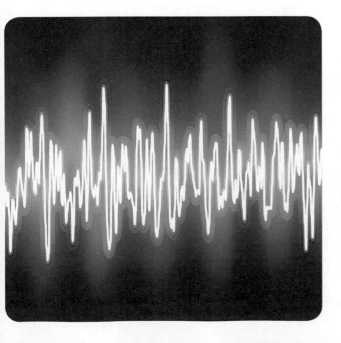

SUMMARY

- Waves can be either transverse or longitudinal.
- Waves are described by their amplitude, wavelength, frequency and period.
- Wave speed can be calculated using the equation,
  wave speed = frequency × wavelength.

QUESTIONS

QUICK TEST

1. Give an example of a longitudinal wave.

2. Give an example of a transverse wave.

3. Define the term 'wavelength'.

4. What are the four properties waves are described by?

EXAM PRACTICE

1. A group of students are investigating waves using a ripple tank and a stroboscope.

   a) What type of waves are they investigating? **[1 mark]**

   b) They record 4 waves passing a point in 1 second.

      Calculate the period of the wave. **[2 marks]**

   c) The wavelength of the waves is 35 cm.

      Calculate the speed of the wave. **[2 marks]**

   d) The students placed a small piece of cork in the tank. They observed it moving up and down as the wave passed but it did not move along the tank.

      Explain the property of waves that this provides evidence of. **[2 marks]**

Electromagnetic waves and properties 1

Electromagnetic waves

Electromagnetic waves are transverse waves that:

● transfer energy from the source of the waves to an absorber

● form a continuous spectrum

● travel at the same velocity through a vacuum (space) or air.

The waves that form the electromagnetic spectrum are grouped in terms of their wavelength and their frequency.

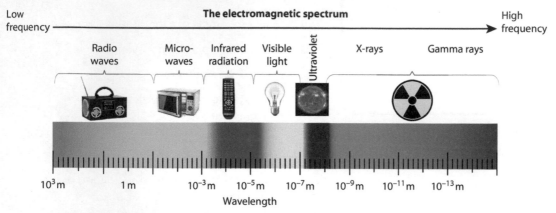

Human eyes only detect visible light so only identify a limited range of electromagnetic radiation.

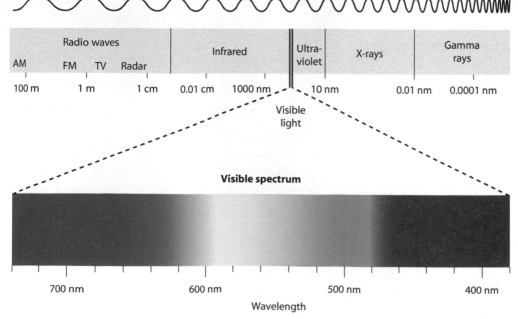

Different wavelengths of electromagnetic waves are reflected, refracted, absorbed or transmitted differently by different substances and types of surface.

Refraction

Refraction is when a wave changes direction as it travels from one medium to another.

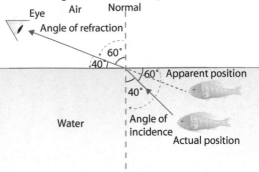

Light refraction through water

In the example above, the fish appears to be above its actual position due to the refraction of the light rays.

HT Refraction occurs due to the wave changing speed as it moves between media. Waves have different velocities in different media.

Light travels faster in the air than it does in the water. This leads to the light ray bending away from the normal. When light slows down, it bends towards the normal.

The wave front diagram below shows a wave moving from air (less optically dense) to water (more optically dense). As the wave travels slower in a denser medium, the edge of the wave that hits the water first slows down whilst the rest of the wave continues at the same speed. This causes the light to bend towards the normal. The opposite effect occurs when a wave moves from a more optically dense medium to a less optically dense medium.

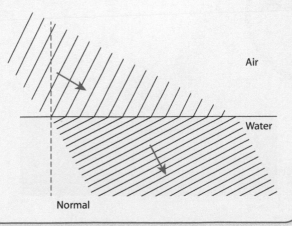

SUMMARY

● **Electromagnetic waves are transverse waves.**

● **The waves that form the electromagnetic spectrum are ordered according to their frequency and wavelength.**

● **Human eyes can only detect visible light.**

● **When a wave changes direction as it travels from one medium to another, it is called refraction.**

QUESTIONS

QUICK TEST

1. What type of electromagnetic waves are used in television remote controls?

2. Which has the highest frequency: visible light or microwaves?

3. Are electromagnetic waves transverse or longitudinal?

4. What is refraction?

HT 5. Which way do waves bend when they enter a more optically dense medium?

EXAM PRACTICE

1. Visible light from the Sun takes the same time to reach the Earth as ultraviolet radiation from the Sun.

   Explain this observation. **[2 marks]**

HT 2. Spear fishing is a traditional means of fishing. The fisherman stands in the water and uses a sharp stick to spear the fish as they swim by.

   Fully explain why the fishermen do not aim the sharp stick exactly where they observe the fish to be. **[3 marks]**

Electromagnetic waves and properties 2

Electromagnetic waves

Changes in atoms and the nuclei of atoms can result in electromagnetic waves being generated or absorbed over a wide frequency range.

Gamma rays originate from changes in the nucleus of an atom.

> **HT Radio waves**
>
> Radio waves can be produced by oscillations in electrical circuits. When radio waves are absorbed they may create an alternating current with the same frequency as the radio wave. Therefore, radio waves induce oscillations in an electrical circuit.

The diagram below shows the main features of electromagnetic waves.

Ultraviolet waves, x-rays and gamma rays can have hazardous effects on human body tissue.

The effects of electromagnetic waves depend on the type of radiation and the size of the dose.

Radiation dose (in sieverts) is a measure of the risk of harm from an exposure of the body to the radiation.

1000 millisieverts (mSv) equals 1 sievert (Sv)

Microwaves can cause internal heating of body cells.

Ultraviolet waves can cause skin to age prematurely and increase the risk of skin cancer.

X-rays and gamma rays are ionising radiation that can cause the mutation of genes and cancer.

Uses of electromagnetic waves

Electromagnetic waves have many practical applications.

Radio waves	Television, radio, bluetooth	Low frequency radio waves can diffract around hills and reflect off the ionosphere, meaning they don't require a direct line of sight between transmitter and receiver. 
Microwaves	Satellite communications, cooking food	Their small wavelength allows them to be directed in narrow beams. HT
Infrared	Electrical heaters, cooking food, infrared cameras	Thermal radiation heats up objects. HT
Visible light	Fibre optic communications	This can be reflected down a fibre optic cable. HT
Ultraviolet	Energy efficient lamps, sun tanning	Ultraviolet is used for low energy light bulbs to produce white light. This requires less energy than filament bulbs. HT
X-rays	Medical imaging and treatments	X-rays are absorbed differently by different parts of the body; more are absorbed by hard tissues, such as bone, and less are absorbed by soft tissues. This allows images of the inside of the body to be created. HT
Gamma rays	Sterilising, medical imaging and treatment of cancer	Gamma rays destroy living cells so can be used to sterilise medical equipment and apparatus. Gamma rays can also be used to destroy cancerous tumours and carry out functional organ scans. HT

SUMMARY

● Electromagnetic waves have many uses, for example radio waves in TV and radio, microwaves for cooking food, and X-rays and gamma rays for medical imaging.

QUESTIONS

QUICK TEST

1. How many millisieverts are in 1 sievert?

HT 2. How are radio waves produced?

3. What are the uses of gamma rays?

4. Give one use of infrared radiation.

EXAM PRACTICE

HT 1. Explain how X-rays are used in medical imaging. **[3 marks]**

2. What are the dangers associated with too much exposure to ultraviolet radiation? **[2 marks]**

Circuits, charge and current

Circuit symbols

The diagram below shows the standard symbols used for components in a circuit.

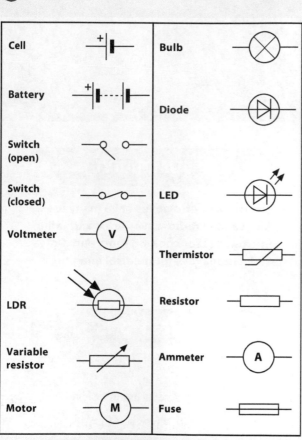

Cell		Bulb	
Battery		Diode	
Switch (open)			
Switch (closed)		LED	
Voltmeter	V		
		Thermistor	
LDR		Resistor	
Variable resistor		Ammeter	A
Motor	M	Fuse	

Here is an example of a circuit diagram:

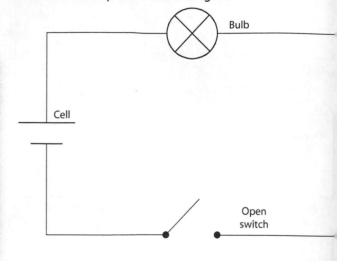

Bulb

Cell

Open switch

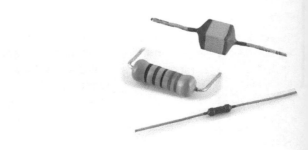

Electrical charge and current

For electrical charge to flow through a closed circuit, the circuit must include a source of energy that produces a potential difference, such as a battery, cell or powerpack.

Electric current is a flow of electrical charge. The size of the electric current is the rate of flow of electrical charge. Charge flow, current and time are linked by the following equation:

> **charge flow = current × time**
>
> $$Q = It$$
>
> - charge flow, Q, in coulombs, C
> - current, I, in amperes or amps A
> - time, t, in seconds, s

Example

A current of 6 A flows through a circuit for 14 seconds. What is the charge flow?

charge flow = current × time

$$= 6 \times 14$$

$$= 84 \text{ C}$$

The current at any point in a single closed loop of a circuit has the same value as the current at any other point in the same closed loop.

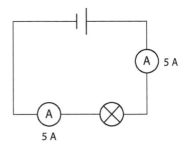

Both ammeters in the circuit above show the same current of 5 amps.

SUMMARY

- Circuit diagrams are drawn using the standard symbols for components in a circuit.
- A source of energy must be present in a closed circuit in order for an electrical charge to flow through the circuit.
- Electrical current is a flow of electrical charge.

QUESTIONS

QUICK TEST

1. What is the symbol for a resistor?

2. What charge flows through a circuit per second if the current is 3.8 A?

3. What is the current of a charge flow of 600 C in three seconds?

EXAM PRACTICE

1. a) A circuit with a single closed loop was being set up.

   For the electrical charge to flow in the circuit, what component must be included in the circuit?

   Explain your answer.   **[2 marks]**

   b) i) The current through a diode in the circuit was 8A and the charge flow was 160 C.

   How long was the current flowing for?   **[2 marks]**

   ii) What would the current be through a fuse in the same circuit?

   Explain your answer.   **[2 marks]**

Current, resistance and potential difference

Current, resistance and potential difference

The current through a component depends on both the resistance of the component and the potential difference (p.d.) across the component. Potential difference is the energy transferred per unit charge passed.

The greater the resistance of the component, the smaller the current for a given potential difference across the component.

Current, potential difference or resistance can be calculated using the following equation:

> **potential difference = current × resistance**
>
> $$V = IR$$
>
> - potential difference, V, in volts, V
> - current, I, in amperes or amps, A
> - resistance, R, in ohms, Ω

> **Example**
>
> A 5 ohm resistor has a current of 2 A flowing through it. What is the potential difference across the resistor?
>
> potential difference = current × resistance
> $$= 5 \times 2$$
> $$= 10 \text{ V}$$

By measuring the current through, and potential difference across a component, it's possible to calculate the resistance of a component.

The circuit diagram (right) would allow you to determine the resistance of the filament lamp.

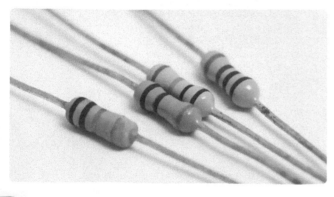

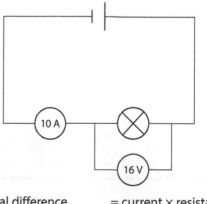

potential difference = current × resistance

resistance $= \dfrac{\text{potential difference}}{\text{current}}$

$$= \frac{16}{10}$$

$$= 1.6 \text{ ohms}$$

Resistors

In an ohmic conductor, at a constant temperature the current is directly proportional to the potential difference across the resistor. This means that the resistance remains constant as the current changes.

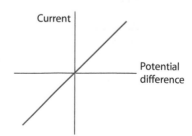

The resistance of components such as lamps, diodes, thermistors and LDRs is not constant; it changes with the current through the component. They are not ohmic conductors.

When a current flows through a resistor, the energy transfer causes the resistor to heat up. This is due to collisions between electrons and the ions in the lattice of the resistor.

This heating can be an advantage, such as in an electrical heater. It is also a disadvantage as it can lead to electrical devices being damaged due to overheating. Thicker wires have a lower resistance as there is a larger cross-sectional area for the current to pass through.

Filament lamps

The resistance of a filament lamp increases as the temperature of the filament increases.

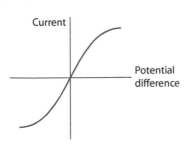

Diodes

The current through a diode flows in one direction only. This means the diode has a very high resistance in the reverse direction.

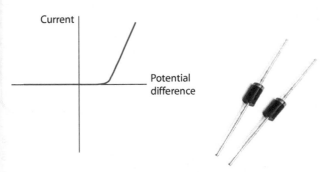

Light dependent resistors (LDR)

The resistance of an LDR decreases as light intensity increases. LDRs are used in circuits where lights are required to switch on when it gets dark, such as floodlights.

Thermistors

The resistance of a thermistor decreases as the temperature increases. Thermistors are used in thermostats to control heating systems.

SUMMARY

- Current, resistance and potential difference are related in that the current through a component depends on both the resistance of the component, and the potential difference across the component.
- Current, potential difference or resistance can be calculated using V = IR.
- In an ohmic conductor, at a constant temperature, the resistance remains constant as the current changes.

QUESTIONS

QUICK TEST

1. Why is an LDR not an ohmic conductor?

2. What is the potential difference if the current is 6 A and the resistance is 3 ohms?

3. Calculate the resistance of a component that has a current of 2 A flowing through it and a potential difference of 8 V.

EXAM PRACTICE

1. An investigation was carried out into the resistance of different components in a circuit which contained a 12V battery pack.

   a) What two pieces of equipment need to be wired into a circuit in order to determine the resistance of a component in the circuit?   **[2 marks]**

   b) A filament lamp is used that has a resistance of 4 ohms.

      What is the current flowing through the filament lamp?   **[2 marks]**

   c) After being left on for a period of time, the resistance of the lamp changed.

      Explain why this occurred and predict how the resistance changed.   **[2 marks]**

Series and parallel circuits

Components can be joined together in either a series circuit or a parallel circuit. Some circuits can include both series and parallel sections.

Series circuits

For components connected in series:

- there is the same current through each component
- the total potential difference of the power supply is shared between the components
- the total resistance of two components is the sum of the resistance of each component.

Total resistance is given by the following equation:

> ### R total = $R1 + R2$
> - resistance, R, in ohms, Ω

Example

What is the total resistance of the two resistors in the series circuit below?

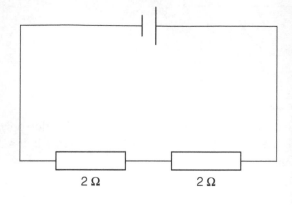

$$\begin{aligned} \text{Total} &= R1 + R2 \\ &= 2\,\Omega + 2\,\Omega \\ &= 4\,\Omega \end{aligned}$$

Parallel circuits

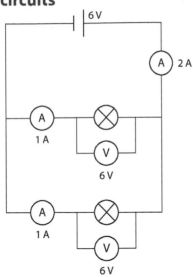

For components connected in parallel:

- the potential difference across each component is the same

- the total current through the whole circuit is the sum of the currents through the separate components. The current splits between the branches of the circuit and combines when the branches meet

- the total resistance of two resistors is less than the resistance of the smallest individual resistor. This is due to the potential difference across the resistors being the same but the current splitting.

SUMMARY

- Components can be joined together in a series circuit or in a parallel circuit.

- In a series circuit, the current is the same through each component; in a parallel circuit, the current through the whole circuit is the total of the currents through the separate components.

QUESTIONS

QUICK TEST

1. How do you calculate the total resistance of two resistors in series?

2. If the current through one component in a series circuit is 4 A, what is the current through the rest of the components?

3. What is a parallel circuit?

EXAM PRACTICE

1. In a parallel circuit a component has a potential difference across it of 9V.

   a) What would be the voltage through the other components in the circuit?

      Explain your answer. **[2 marks]**

   b) Each component has a resistance of 3 ohms.

      What conclusion can be made about the total resistance of all the components in the circuit?

      Explain your answer. **[2 marks]**

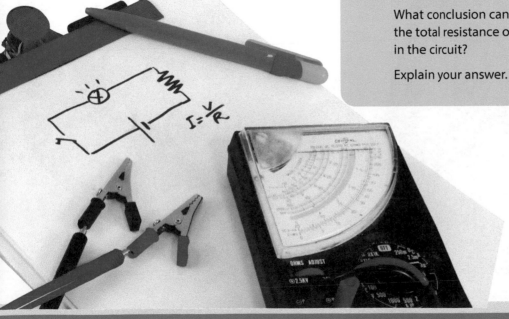

Domestic uses and safety

Direct and alternating current

Cells and batteries supply current that always passes in the same direction. This is direct current (dc).

Alternating current (ac) changes direction at a frequency of fifty times a second. Mains electricity is an ac supply. In the UK it has a frequency of 50 Hz and is about 230 V.

Mains electricity

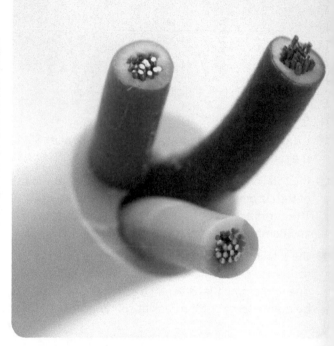

WS Most electrical appliances are connected to the mains using a three-core cable with a three-pin plug.

Live wire	Brown	Carries the alternating potential difference from the supply.
Neutral wire	Blue	Completes the circuit. The neutral wire is at, or close to, earth potential (0 V).
Earth wire	Green and yellow stripes	The earth wire is at 0 V. It only carries a current if there is a fault.

The potential difference between the live wire and earth (0 V) is about 230 V.

Our bodies are at earth potential (0 V). Touching a live wire produces a large potential difference across our body. This causes a current to flow through our body, resulting in an electric shock that could cause serious injury or death.

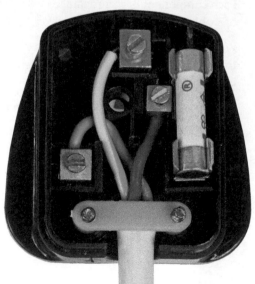

Insulation, fuses and circuit breakers

If an electrical fault causes too great a current, the circuit is disconnected by a fuse or a circuit breaker connected to the live wire.

The current will cause the fuse to overheat and melt or the circuit breaker to switch off (trip). A circuit breaker operates much faster than a fuse and can be reset.

Appliances with metal cases are usually earthed. If a fault occurs, a large current flows from the live wire to earth. This melts the fuse and disconnects the live wire.

Some appliances are double insulated meaning it is impossible for the case to become live. (Either the case is plastic or it is impossible for the live wire to come into contact with the casing). Double insulated appliances have no earth connection.

Electric drills are examples of appliances that are double insulated.

Circuit breaker

Fuses

SUMMARY

- Current that always flows in the same direction is called direct current (dc).
- Current that changes direction is called alternating current (ac). Mains electricity is ac.
- Most electrical appliances use a plug to connect to mains electricity. A plug has a live wire, neutral wire and earth wire.
- A fuse or circuit breaker disconnects the circuit if the current is too great.

QUESTIONS

QUICK TEST

1. What do ac and dc stand for?
2. What are the advantages of a circuit breaker over a conventional fuse?
3. What colour is the live wire in a plug?
4. What voltage does the earth wire have in a plug?

EXAM PRACTICE

1. A student was investigating the plugs of different electrical appliances.

   He was examining the plug of an electric appliance which had a plastic case. He noticed that this plug did not have an earth wire.

   He wrote in his report that all appliances must have an earth wire and this particular appliance was therefore unsafe to use.

   a) Was the student correct?

      Explain your answer. [4 marks]

   b) Explain why a fuse is important in an appliance with an earth wire. [2 marks]

Energy transfers

Power

The power of a device is related to the potential difference across it and the current through it by the following equations:

power = potential difference × current

$$P = VI$$

or

power = current² × resistance

$$P = I^2 R$$

- power, P, in watts, W
- potential difference, V, in volts, V
- current, I, in amperes or amps, A
- resistance, R, in ohms, Ω

Example

A bulb has a potential difference of 240 V and a current flowing through it of 0.6 A. What is the power of the bulb?

power = potential difference × current

$$= 240 \times 0.6$$

$$= 144 \, W$$

Energy transfers in everyday appliances

Everyday electrical appliances are designed to bring about energy transfers.

The amount of energy an appliance transfers depends on how long the appliance is switched on for and the power of the appliance.

Here are some examples of everyday energy transfer in appliances:

- A hairdryer transfers electrical energy from the ac mains to kinetic energy (in an electric motor to drive a fan) and heat energy (in a heating element).

- A torch transfers electrical energy from batteries into light energy from a bulb.

Work done

Work is done when charge flows in a circuit.

The amount of energy transferred by electrical work can be calculated using the following equation:

> **energy transferred = power × time**
>
> $$E = Pt$$
>
> and
> **energy transferred = charge flow × potential difference**
>
> $$E = QV$$
>
> - energy transferred, E, in joules, J
> - power, P, in watts, W
> - time, t, in seconds, s,
> - charge flow, Q, in coulombs, C
> - potential difference, V, in volts, V

The National Grid

The National Grid is a system of cables and transformers linking power stations to consumers.

Electrical power is transferred from power stations to consumers using the National Grid.

Step-up transformers **increase** the potential difference from the power station to the transmission cables.

Step-down transformers **decrease** the potential difference to a much lower and safer level for domestic use.

Increasing the potential difference reduces the current so reduces the energy loss due to heating in the transmission cables. Reducing the loss of energy through heat makes the transfer of energy much more efficient. Also, the wires would glow and be more likely to break over time if the current through them was high.

SUMMARY

- Power can be calculated using the equation: **power = potential difference × current** or **power = current² × resistance**.
- Electrical appliances bring about energy transfers, for example, a torch transfers electrical energy into light energy.
- The National Grid is a system for getting electricity to consumers. Step-up transformers increase potential difference from the power station to the cables; step-down transformers decrease potential difference to a safe level for homes.

QUESTIONS

QUICK TEST

1. Explain why step-down transformers are important in the National Grid.

2. What is the energy transferred by a charge flow of 50 C and a potential difference of 10 V?

3. What is the power of a device that has a current flowing through it of 4 A and a resistance of 3 Ω?

EXAM PRACTICE

1. An overhead cable has a current of 500 A and a power of 8000 kW.

   a) What is the resistance of the cable? **[2 marks]**

   b) What is energy transferred by the cables in 120 seconds?

      Give your answer in kJ. **[2 marks]**

   c) Explain the advantage of keeping the current relatively low in these wires. **[2 marks]**

Permanent and induced magnetism, magnetic forces and fields

Poles of a magnet

The poles of a magnet are the places where the magnetic forces are strongest.

When two magnets are brought close together they exert a force on each other.

> Two like poles repel. Two unlike poles attract.

Attraction between opposite poles

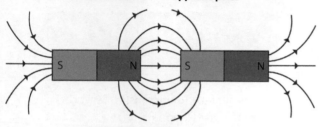

Repulsion between like poles

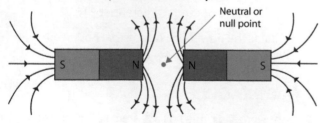

 Neutral or null point

Magnetism is an example of a non-contact force.

Permanent magnetism vs induced magnetism

A permanent magnet . . .

● produces its own magnetic field.

An induced magnet . . .

● becomes a magnet when placed in a magnetic field

● always experiences a force of attraction

● loses most or all of its magnetism quickly when removed from a magnetic field.

Magnetic field

The region around a magnet – where a force acts on another magnet or on a magnetic material (iron, steel, cobalt, magnadur and nickel) – is called the magnetic field.

WS

A compass can be used to plot a magnetic field

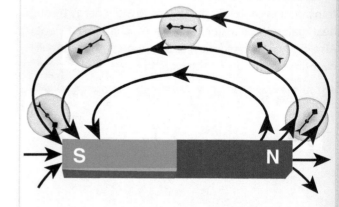

The **force** between a magnet and a magnetic material is always attraction.

The **strength** of the magnetic field depends on the distance from the magnet.

The **field** is strongest at the poles of the magnet.

The **direction** of a magnetic field line is from the north pole of the magnet to the south pole of the magnet.

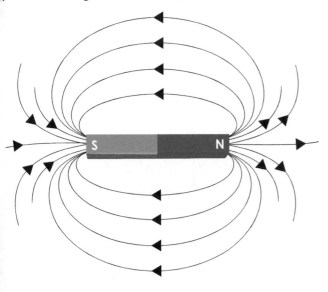

Compasses

A magnetic compass contains a bar magnet that points towards magnetic north. This provides evidence that the Earth's core is magnetic and produces a magnetic field.

SUMMARY

● Opposite poles of a magnet attract each other.

● Like poles of a magnet repel each other.

● A permanent magnet produces its own magnetic field; an induced magnet becomes a magnet when placed in a magnetic field.

● A magnetic field is the area around a magnet.

QUESTIONS

QUICK TEST

1. What happens when the two south poles of a bar magnet are brought together?

2. What happens if opposite poles of two magnets are brought together?

3. What type of force is magnetism?

4. Why does a magnetic compass point north?

EXAM PRACTICE

1. A student was carrying out an investigation into the magnetic field around a bar magnet.

   a) What is a magnetic field? [1 mark]

   b) Predict where the magnetic field strength would be the strongest. [1 mark]

   c) What force would a magnetic material experience when inside the magnetic field? [1 mark]

   d) The student used a compass to plot the magnetic field lines, in what direction would the field lines run? [1 mark]

Electromagnets, Fleming's left-hand rule and electric motors

Electromagnets

When a current flows through a conducting wire a magnetic field is produced around the wire.

The shape of the magnetic field can be seen as a series of concentric circles in a plane, perpendicular to the wire.

The direction of these field lines depends on the direction of the current.

The strength of the magnetic field depends on the current through the wire and the distance from the wire.

Coiling the wire into a solenoid (a helix) increases the strength of the magnetic field created by a current through the wire.

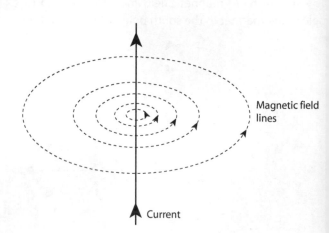

Magnetic field lines

Current

Magnetic field has a similar shape to that of a bar magnet.

Adding an iron core increases the magnetic field strength of a solenoid.

The fields from individual coils in the solenoid add together to form a very strong, almost uniform, field along the centre of the solenoid.

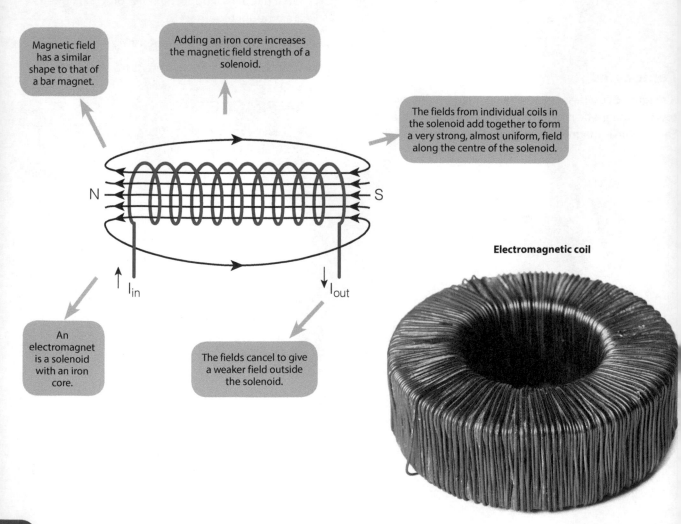

N   S

I_{in} I_{out}

An electromagnet is a solenoid with an iron core.

The fields cancel to give a weaker field outside the solenoid.

Electromagnetic coil

🅗 Fleming's left-hand rule and the motor effect

When a conductor carrying a current is placed in a magnetic field, the magnet producing the field and the conductor exert an equal and opposite force on each other. This is the motor effect and is due to interactions between magnetic fields.

The direction of the force on the conductor can be identified using Fleming's left-hand rule.

If the direction of the current or the direction of the magnetic field is reversed, the direction of the force on the conductor is reversed.

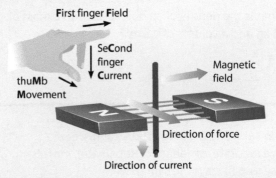

The size of the force on the conductor depends on:

● the magnetic flux density

● the current in the conductor

● the length of conductor in the magnetic field.

For a conductor at right angles to a magnetic field and carrying a current, the force can be calculated using the following equation:

---
force = magnetic flux density × current × length

$$F = BIl$$

● force, F, in newtons, N
● magnetic flux density, B, in tesla, T
● current, I, in amperes, A (or amp)
● length, l, in metres, m

Example

What is the force produced by a 0.5 m long conductor, with a magnetic flux of 1.2 T and a current of 16 A flowing through it?

$F = BIl$

$F = 1.2 \times 16 \times 0.5 = 9.6$ N

🅗 Electric motors

A coil of wire carrying a current in a magnetic field experiences a force, causing it to rotate. This is the basis of an electric motor.

The commutator and graphite brush allow the current to be reversed every half turn to keep the coil spinning.

Simple electric motor

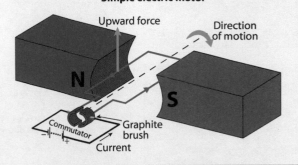

SUMMARY

● When a current flows through a conducting wire, a magnetic field is produced around the wire.

● To increase the strength of the magnetic field, the wire can be coiled into a solenoid.

● 🅗 An electric motor can be created by a coil of wire carrying a current in a magnetic field, which causes it to rotate.

QUESTIONS

QUICK TEST

1. Name the two things that the strength of a magnetic field around a wire depends on.

🅗 2. What three variables are related by Fleming's left-hand rule?

EXAM PRACTICE

🅗 1. Calculate the force produced by an 8 A current running through a 9 m wire which has a magnetic flux density of 3.2 T.　**[2 marks]**

The particle model and pressure

The particle model

Matter can exist as a solid, liquid or as a gas.

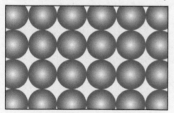

Solid – particles are very close together and vibrating. They are in fixed positions.

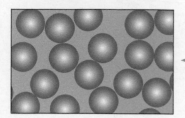

Liquid – particles are very close together but are free to move relative to each other. This allows liquids to flow.

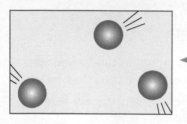

Gas – particles in a gas are not close together. The particles move rapidly in all directions.

If the particles in a substance are more closely packed together, the density of the substance is higher. This means that liquids have a higher density than gases. Most solids have a higher density than liquids.

Density also increases when the particles are forced into a smaller volume.

Low density

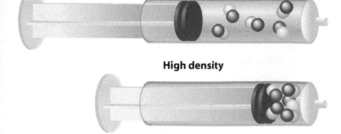
High density

The density of a material is defined by the following equation:

$$\text{density} = \frac{\text{mass}}{\text{volume}}$$
$$p = \frac{m}{v}$$

- density, ρ, in kilograms per metre cubed, kg/m³
- mass, m, in kilograms, kg
- volume, V, in metres cubed, m³

Example

What is the density of an object that has a mass of 56 kg and a volume of 0.5 m³?

$$\rho = \frac{m}{v}$$
$$= \frac{56}{0.5} = 112 \text{ kg/m}^3$$

When substances change state (melt, freeze, boil, evaporate, condense or sublimate), mass is conserved (it stays the same).

Changes of state are physical changes: the change does not produce a new substance, so if the change is reversed the substance recovers its original properties.

Ice

Water

Steam

Gas under pressure

The molecules of a gas are in constant random motion.

When the molecules collide with the wall of their container they exert a force on the wall. The total force exerted by all of the molecules inside the container on a unit area of the wall is the gas pressure.

Increasing the temperature of a gas, held at constant volume, increases the pressure exerted by the gas.

Decreasing the temperature of a gas, held at constant volume, decreases the pressure exerted by the gas.

The temperature of the gas is related to the average kinetic energy of the molecules. The higher the temperature, the greater the average kinetic energy, and so the faster the average speed of the molecules. At higher temperatures, the particles collide with the walls of the container at a higher speed.

SUMMARY

- Matter can exist as solid, liquid or gas.
- Gas pressure is the force exerted by all the molecules inside a container on a unit area of the wall.
- The temperature of a gas depends on the average kinetic energy of the molecules.

QUESTIONS

QUICK TEST

1. In what three states can matter exist?

2. What is the density of a 4 kg object with a volume of 0.002 m^3?

3. What happens to the mass of a liquid when it freezes?

4. In a gas, particles are not close together. True or false?

EXAM PRACTICE

1. A scientist carried out an investigation into the effect of temperature of gases.

   a) The scientist heated up the gas.

      Explain the effect this would have on the pressure of the gas. **[1 mark]**

   b) What variable would be important to keep constant during this investigation? **[1 mark]**

   c) The scientist cooled the gas to a temperature lower than its boiling point.

      What effect would this have on the gas? **[1 mark]**

   d) The scientist heated the substance and the gas regained its original properties.

      Explain why this was possible. **[2 marks]**

Internal energy and change of state

Internal energy

Energy is stored inside a system by the particles (atoms and molecules) that make up the system. This is called **internal energy**.

Internal energy of a system is equal to the total kinetic energy and potential energy of all the atoms and molecules that make up the system.

Heating changes the energy stored within the system by increasing the energy of the particles that make up the system. This either raises the temperature of the system or produces a change of state.

Heat and temperature are related but are not a measure of the same thing.

- Heat is the amount of thermal energy and is measured in Joules (J).

- Temperature is how hot or cold something is and is measured in degrees Celsius (°C).

Changes of state and specific latent heat

When a change of state occurs, the stored internal energy changes, but the temperature remains constant. The graph below shows the change in temperature of water as it is heated; the temperature is constant when the water is changing state.

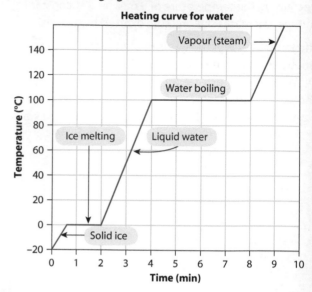

Heating curve for water

The specific latent heat of a substance is equal to the energy required to change the state of one kilogram of the substance with no change in temperature.

The energy required to cause a change of state can be calculated by the following equation:

$$\text{energy for a change of state} = \text{mass} \times \text{specific latent heat}$$
$$E = mL$$

- energy, E, in joules , J
- mass, m, in kilograms, kg
- specific latent heat, L, in joules per kilogram, J/kg

Example

What is the energy needed for 600 g of water to melt? (The specific latent heat of water melting is 334 kJ/kg.)

$$E = mL$$
$$0.6 \times 334 = 200.4 \text{ kJ}$$

The specific latent heat of fusion is the energy required for a change of state from solid to liquid.

The specific latent heat of vapourisation is the energy required for a change of state from liquid to vapour.

Temperature can be measured in degrees Celsius (°C) or Kelvin (K). To convert from Celsius to Kelvin, add 273.

For example:

- 10°C + 273 = 283 K, so 10°C is equal to 283 K.

0 Kelvin (−273°C) is **absolute zero**. At this point, the particles have no kinetic energy so are not moving.

SUMMARY

- Specific latent heat of a substance is equal to the energy required to change the state of 1 kg of the substance with no change in temperature.
- Internal energy is the energy stored inside a system by the particles that make up the system.
- Heat is thermal energy and is measured in Joules.
- Temperature is how hot or cold something is and is measured in degrees Celsius or Kelvin.

QUESTIONS

QUICK TEST

1. What energy is needed for 300 kg of cast iron to turn from a solid to a liquid? (The specific latent heat of iron melting is 126 kJ/kg.)

2. What is 28 °C in Kelvin?

3. What is 290 Kelvin in degrees Celsius?

EXAM PRACTICE

1. An investigation was carried out into the heating of a substance.

   a) Explain what happens to the energy stored within a system when it is heated. **[1 mark]**

   b) Describe how the internal energy of a system relates to the kinetic energy and potential energy of the atoms which make up the system. **[2 marks]**

   c) At one point in the investigation the substance was being heated but its temperature was not increasing.

   Explain this observation. **[1 mark]**

Atoms and isotopes

Structure of atoms

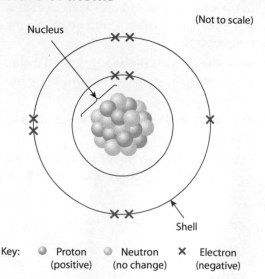

(Not to scale)

Nucleus

Shell

Key: ● Proton (positive)   ● Neutron (no change)   ✕ Electron (negative)

Atoms have a radius of around 1×10^{-10} metres.

The radius of a nucleus is less than $\frac{1}{10\,000}$ of the radius of an atom.

Most of the mass of an atom is concentrated in the nucleus. Protons and neutrons have a relative mass of 1 while electrons have a relative mass of 0.0005. The electrons are arranged at different distances from the nucleus (are at different energy levels).

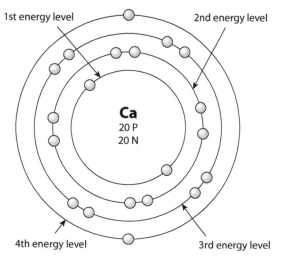

1st energy level

2nd energy level

Ca
20 P
20 N

4th energy level

3rd energy level

Absorption of electromagnetic radiation causes the electrons to become excited and move to a higher energy level and further from the nucleus.

Emission of electromagnetic radiation causes the electrons to move to a lower energy level and move closer to the nucleus.

If an atom loses or gains an electron, it is ionised.

The number of electrons is equal to the number of protons in the nucleus of an atom.

Atoms have no overall electrical charge.

All atoms of a particular element have the same number of protons. The number of protons in an atom of an element is called the **atomic number**.

The total number of protons and neutrons in an atom is called the **mass number**.

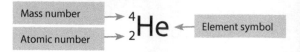

Mass number → $^{4}_{2}\text{He}$ ← Element symbol
Atomic number →

Atoms of the same element can have different numbers of neutrons; these atoms are called **isotopes** of that element. For example, below are some isotopes of nitrogen. They each have 7 protons in the nucleus but different numbers of neutrons, giving the different isotopes.

$$^{14}\text{N} \quad ^{15}\text{N} \quad ^{13}\text{N}$$

Atoms turn into **positive ions** if they lose one or more outer electrons and into **negative ions** if they gain one or more outer electrons.

WS The development of the atomic model

Before the discovery of the electron, atoms were thought to be tiny spheres that could not be divided.

⬇

The discovery of the electron led to further developments of the model. The plum pudding model suggested that the atom is a ball of positive charge with negative electrons embedded in it.

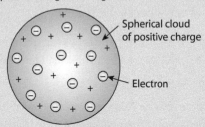

Spherical cloud of positive charge

Electron

⬇

Rutherford, Geiger and Marsden's alpha scattering experiment led to the conclusion that the mass of an atom was concentrated at the centre (nucleus) and that the nucleus was charged.

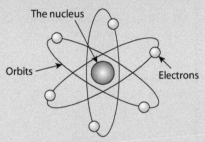

The nucleus

Orbits

Electrons

⬇

This evidence led to the nuclear model replacing the plum pudding model.

⬇

Niels Bohr suggested that the electrons orbit the nucleus at specific distances. The theoretical calculations of Bohr agreed with experimental observation.

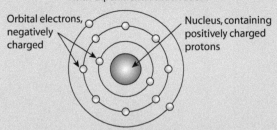

Orbital electrons, negatively charged

Nucleus, containing positively charged protons

⬇

Later experiments led to the idea of the nucleus containing smaller particles with the same amount of positive charge (protons).

⬇

In 1932, the experimental work of James Chadwick provided evidence of the existence within the nucleus of the neutron.

SUMMARY

- Atoms are made up of protons, neutrons and electrons.
- If an atom gains or loses an electron, it is ionised.
- There were many stages and a number of scientists involved in the development of the atomic model.

QUESTIONS

QUICK TEST

1. What charges do protons, neutrons and electrons have?

2. What name is given to atoms of the same element that have different numbers of neutrons?

3. Atoms turn into negative ions if they lose outer electrons. True or false?

EXAM PRACTICE

1. a) Complete the table below to show the relative masses of protons, neutrons and electrons. **[2 marks]**

Particle	Relative mass
Proton	
Neutron	
Electron	

b) Compare the charge of the nucleus with the charge of the atom overall. **[2 marks]**

c) What happens to the electrons of an atom when the atom absorbs electromagnetic radiation? **[2 marks]**

Radioactive decay, nuclear radiation and nuclear equations

Radioactive decay

Some atomic nuclei are unstable. The nucleus gives out radiation as it changes to become more stable. This is a random process called radioactive decay. Changes in atoms and nuclei can also generate and absorb radiation over the whole frequency range.

Activity is the rate at which a source of unstable nuclei decays, measured in becquerel (Bq).

● 1 becquerel = 1 decay per second

Count rate is the number of decays recorded each second by a detector, such as a Geiger-Müller tube.

● 1 becquerel = 1 count per second

Radioactive decay can release a neutron, alpha particles, beta particles or gamma rays. If radiation is ionising, it can damage materials and living cells.

Particle	Description	Penetration in air	Absorbed by...	Ionising power
Alpha particles (α)	Two neutrons and two protons (a helium nucleus).	a few centimetres	a thin sheet of paper	strongly ionising
Beta particles (β)	High speed electron ejected from the nucleus as a neutron turns into a proton.	a few metres	a sheet of aluminium about 5 mm thick	moderately ionising
Gamma rays (γ)	Electromagnetic radiation from the nucleus.	a large distance	a thick sheet of lead or several metres of concrete	weakly ionising

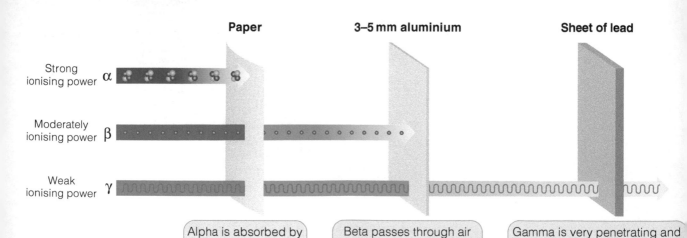

Alpha is absorbed by a few centimetres of air or a thin sheet of paper.

Beta passes through air and paper but is absorbed by a few millimetres of aluminium.

Gamma is very penetrating and needs many centimetres of lead or many metres of concrete to absorb most of it.

Nuclear equations

Nuclear equations are used to represent radioactive decay.

Nuclear equations can use the following symbols:

$$^{4}_{2}\text{He} \quad \text{alpha particle}$$

$$^{0}_{-1}\text{e} \quad \text{beta particle}$$

Alpha decay causes both the mass and charge of the nucleus to decrease, as two protons and two neutrons are released.

$$^{219}_{86}\text{radon} \longrightarrow ^{215}_{84}\text{polonium} + ^{4}_{2}\text{He}$$

Beta decay does not cause the mass of the nucleus to change but does cause the charge of the nucleus to change, as a proton becomes a neutron.

$$^{14}_{6}\text{carbon} \longrightarrow ^{14}_{7}\text{nitrogen} + ^{0}_{-1}\text{e}$$

The above example is β– decay as a neutron has becomes a proton and an electron has been ejected. In β+ decay a proton becomes a neutron plus a positron.

The emission of a gamma ray does not cause the mass or the charge of the nucleus to change.

Alpha decay

Beta-minus decay with gamma ray

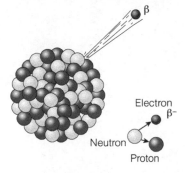

Electron
β–

Neutron

Proton

SUMMARY

- Radioactive decay occurs when unstable atomic nuclei give out radiation as they change to become stable.
- The rate at which radioactive decay occurs is measured in Becquerels.
- Radioactive decay can release a neutron, alpha particles, beta particles or gamma rays.
- Nuclear equations represent radioactive decay.

QUESTIONS

QUICK TEST

1. How far does beta radiation penetrate in air?

2. What material is required to absorb alpha particles?

3. What effect does beta decay have on the mass and charge of the nucleus of an atom?

EXAM PRACTICE

1. The equation below shows an example of radioactive decay.

$$^{149}_{A}\text{Gd} \rightarrow ^{B}_{62}\text{Sm} + ^{4}_{2}\text{He}$$

   a) Identify this type of decay. **[1 mark]**

   b) Give the values of the missing mass numbers A and B. **[2 marks]**

   c) Compare the effect of this type of decay on the mass and charge of the nucleus with the effect of gamma ray emission. **[2 marks]**

Half-lives and the random nature of radioactive decay

Half-life

Radioactive decay occurs randomly. It is not possible to predict which nuclei will decay.

The half-life of a radioactive isotope is the average time it takes for:

● the number of nuclei in a sample of the isotope to halve

or

● the count rate (or activity) from a sample containing the isotope to fall to half of its initial level.

Uranium

Example

A radioactive sample has an activity of 560 counts per second. After 8 days, the activity is 280 counts per second. This gives a half-life of 8 days.

The decay can be plotted on a graph and the half-life determined from the graph.

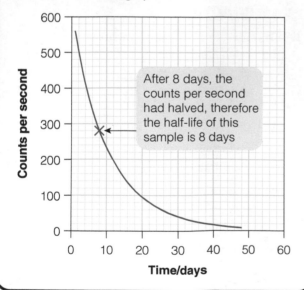

After 8 days, the counts per second had halved, therefore the half-life of this sample is 8 days

HT After another 8 days, the activity would now be 140 counts per second. Therefore, in two half-lives the activity has declined by a total of 560 − 140 = 420 counts per second. As a ratio, the net decline is 420 : 560, or 3 : 4.

Radioactive contamination

Radioactive contamination is the unwanted presence of materials containing radioactive atoms or other materials.

This is a hazard due to the decay of the contaminating atoms. The level of the hazard depends on the type of radiation emitted.

Irradiation is the process of exposing an object to nuclear radiation. This is different from radioactive contamination as the irradiated object does not become radioactive.

Suitable precautions must be taken to protect against any hazard from the radioactive source used in the process of irradiation. In medical testing using radioactive sources, the doses patients receive are limited and medical staff wear protective equipment.

It is important for the findings of studies into the effects of radiation on humans to be published and shared with other scientists. This allows the findings of the studies to be checked by other scientists by the peer review process.

SUMMARY

- Half-life of a radioactive isotope is the average time it takes for the number of nuclei in a sample of the isotope to halve, or the activity from a sample of the isotope to fall to half of its initial level.
- Materials that contain unwanted radioactive atoms or other materials are referred to as radioactive contamination.
- Exposing a material to nuclear radiation is called irradiation.

QUESTIONS

QUICK TEST

1. What is irradiation?

2. Why is radioactive contamination a hazard?

EXAM PRACTICE

1. Researchers carried out an investigation on a radioactive sample which has an activity of 3178 counts per second.

   a) What value will the count rate be after one half life?

      Show how you arrived at your answer. **[2 marks]**

   b) The half life was 6 hours. What will be the count rate after 1 day?

      Show how you arrived at your answer. **[2 marks]**

   c) i) During an investigation into this sample some of the material was found on researchers' clothes.

      Is this an example of contamination or irradiation?

      Explain your answer. **[2 marks]**

   ii) Why is it important that researchers publish the results of their findings on this sample? **[2 marks]**

Answers

Biology

Page 7
QUICK TEST
1. False. Eukaryotic cells are more complex than prokaryotic cells.
2. False – most cells do have a nucleus, but others, e.g. red blood cells, bacterial cells, do not.
3. Ribosomes

EXAM PRACTICE
1. a) Plasmids [1]
   b) They possess a cell wall. [1]
   c) Muscle cells release a lot of energy [1] mitochondria release this energy (for contraction) [1]
   d) $10/500 = 0.002 \times 1000$
      $= 2\ \mu m$ [2 marks for correct answer. If incorrect, 1 mark for correct working.]

Page 9
QUICK TEST
1. Ciliated epithelial cells
2. Xylem
3. Meristems

EXAM PRACTICE
1. Differentiation [1]
2. a) One from: Long, slender axons for carrying impulses long distances; Many dendrites for connecting with other nerve cells [1]
   b) Two from: Therapeutic cloning; Treating paralysis; Repairing nerve damage; Cancer research; Growing new organs for transplantation [2]
   c) Two from: Stem cells are sometimes obtained from human embryos; They believe it is wrong to use embryos (which might be deemed living) for these purposes; Risk of viral infections [2]

Page 11
QUICK TEST
1. Light microscope; electron microscope
2. To stain the nuclei of animal cells
3. False. Light microscopes can only produce 2D images.

EXAM PRACTICE
1. 82.61 mm^2 [3]
   [If incorrect, 1 mark each for correct working: Mean radius = 5.13 mm; area = 3.14 x (5.13)^2]
2. a) Objective lenses [1]
   b) Mitochondria are very small [1] Resolving power of light microscope not great enough [1]

c) Scanning electron microscope (SEM) [1] Light microscopes (and transmission electron microscopes) cannot produce 3D images [1]

Page 13
QUICK TEST
1. Gametes
2. Chromosomes
3. DNA/chromosomes and all other organelles

EXAM PRACTICE
1. a) Meiosis [1]
   b) Four cells produced (in second meiotic division) [1] Each cell is haploid (not diploid) [1]

Page 15
QUICK TEST
1. Aerobic
2. A reaction that gives out heat
3. Lactic acid
4. A catabolic reaction

EXAM PRACTICE
1. a) Glucose → Lactic acid + energy released [1] Glucose → Carbon dioxide + Ethanol + Energy released [1]
   b) 1 mark must be one from: Ethanol excreted/removed from cell; But cannot be removed from surroundings; Humans sometimes extract it for the alcoholic drinks industry [1]
      Two from: Lactic acid causes oxygen debt; Increased breathing and heart rates deliver more blood to muscles; Oxygen removes lactic acid; Lactic acid oxidised [2]

Page 17
QUICK TEST
1. pH and temperature
2. Lipases
3. Amino acids

EXAM PRACTICE
1. a) Three from: High temperature denatures enzyme; Active site shape changed; Starch no longer fits active site; No maltose formed [3]
   b) Three from: Acidic pH keeps active site in correct shape; Protein substrate fits active site; Enzyme puts a strain on protein molecule/breaks it down; To products/peptides/amino acids [3]

Page 19

QUICK TEST

1. To calculate rates of diffusion
2. The plant cells have a higher water potential/lower solute concentration. The water moves down an osmotic gradient from inside the cells into the salt solution by osmosis.

EXAM PRACTICE

1. **Two from:** Ions enter by active transport; requiring release of energy; from respiration; via protein carriers in the cell membrane **[2]**
2. **Four from:** The cell has become plasmolysed; The outside solution has lower water potential/higher solute concentration than cell contents; Water leaves cell; By osmosis; Down concentration/water potential/osmotic gradient **[4]**

Page 21

QUICK TEST

1. The leaf
2. Lignin

EXAM PRACTICE

1. a) **One from:** Xylem consist of hollow tubes while phloem have sieve end-plates; Xylem vessels do not possess companion cells. **[1]**
   b) Phloem carries sugar/sucrose **[1]**
      Aphids need sugar for diet/have well-adapted mouthparts to reach source of sugar. **[1]**

Page 23

QUICK TEST

1. It would increase the (evapo)transpiration rate.
2. Guard cells
3. Potometer

EXAM PRACTICE

1. a) 1st 60 mins: Bubble moves 1.2 cm (2.3 – 2.1) **[1]** therefore rate is 1.2 / 60 = 0.02 cm per min. **[1]**
      2nd 60 mins: Bubble moves 3.5 cm **[1]**
      Therefore rate is 3.4 / 60 = 0.06 cm per min **[1]**
      [Correct answer in each case gains full 2 marks automatically.]
   b) **Three from:** Moving air increases rate of water absorption; Water vapour lost more rapidly from leaves/faster transpiration; Moving air increases diffusion gradient between leaf interior and atmosphere; Evaporation/diffusion is therefore more rapid **[3]**

Page 25

QUICK TEST

1. Veins
2. True
3. The pulmonary vein.

EXAM PRACTICE

1. a) Artery has to withstand/recoil with higher pressure, elasticity allows smoother blood flow/second boost to blood when recoils **[1]**
   b) Vein has valves to prevent backflow of blood/compensate for low blood pressure **[1]**
2. a) The blood passes through the heart twice for every complete circuit of the body. **[1]**
   b) **One from:** Allows a higher pressure to be maintained around the body; Allows maximum efficiency for delivering oxygen/materials to body cells. **[1]**

Page 27

QUICK TEST

1. Red blood cells, white blood cells, plasma and platelets
2. Oxyhaemoglobin

EXAM PRACTICE

1. **Two from:** Involved in defence against infection/part of immune system; Some engulf foreign cells; Others produce antibodies **[2]**
2. **Three from:** Tar layer increases diffusion path for oxygen; Less oxygen is absorbed into the bloodstream; Less oxygen is delivered to respiring cells; Breathing rate is increased to compensate for this **[3]**

Page 29

QUICK TEST

1. Photosynthesis takes **in** carbon dioxide and water, and requires an energy input. It produces glucose and oxygen. Respiration **produces** carbon dioxide and water, and releases energy. It absorbs glucose and oxygen.
2. It is needed for cell walls.

EXAM PRACTICE

1. a) Increased light intensity increases rate of photosynthesis **[1]** More photosynthesis means more carbohydrate produced in a given time **[1]** More carbohydrate means bigger/more tomatoes **[1]** She should install more powerful lighting/artificial lighting at night **[1]**
   b) **Two from:** Rate of photosynthesis does not increase after a certain light intensity; Therefore any extra investment in lighting will not result in any more yield; Money will be wasted **[2]**

Answers

QUICK TEST

1. Body mass index
2. A microorganism that causes disease

EXAM PRACTICE

1. **Two from:** Emphysema; Heart disease; Stroke; **Any other reasonable answer. [2]**
2. a) 26.7 **[2] [If incorrect, allow 1 mark for correct working:]** $64/1.55^2$
   b) She needs to lose some weight **[1]** Her BMI category is 'overweight.' **[1]**

Page 33
QUICK TEST

1. Cholera bacteria are found in human faeces, which contaminate water supplies.
2. **Any reasonable answer, e.g.:** athlete's foot

EXAM PRACTICE

1. a) Pathogens **[1]**
   b) **Two from:** Bacteria and viruses reproduce rapidly in the body; Viruses cause cell damage; Toxins are produced that damage tissues **[2]**
2. Warm temperatures are ideal for mosquitoes to thrive **[1]** Stagnant water is ideal habitat for mosquito eggs to be laid/larvae to survive. **[1]**

Page 35
QUICK TEST

1. Phagocyte engulfs a pathogen then digests it using the cell's enzymes.
2. An antigen is a molecular marker on a pathogen cell membrane that acts as a recognition point for antibodies.
3. Lysozyme

EXAM PRACTICE

1. a) Antibodies **[1]**
   b) i) Lymphocyte **[1]**
      ii) Antigen is identified by the immune system **[1]** Memory cells are activated **[1]** A large amount of antibodies is produced rapidly to combat the fresh invasion of pathogens. **[1]**

Page 37
QUICK TEST

1. An antibiotic is a drug that kills bacteria.
2. Active and passive
3. Do not prescribe antibiotics for viral/non-serious infections; Ensure that the full course of antibiotics is completed.

EXAM PRACTICE

1. Dead/inactive form of virus is injected into person **[1]** this triggers production of antibodies **[1]** and formation of memory cells **[1]** If actual infection occurs, antibodies produced rapidly to deal with the pathogen **[1]**
2. a) Inhibit cell processes in bacteria **[1]**
   b) **Two from:** Antibiotics do not work against viruses/only work against bacteria; Flu caused by a virus; Over prescription risks antibiotic resistance **[2]**
3. MRSA is resistant to most modern antibiotics **[1]** caused by doctors over-prescribing antibiotics **[1]** new antibiotics need to be developed, which is a slow process/takes many years. **[1]**

Page 39
QUICK TEST

1. A blind trial is when volunteers don't know whether they have been given the new drug or a placebo.
2. True

EXAM PRACTICE

1. a) **Two from:** To ensure that the drug is actually effective/more effective than placebo; To work out the most effective dose/method of application; To comply with legislation **[2]**
   b) **One from:** Computer modelling; Experimenting on cell cultures **[1]**
2. a) Therapeutic ratio = 100 / 5
                         = 20

   [2 for correct answer. If answer is incorrect, 1 mark for correct working.]
   b) 20 is a bigger value than 6 (for morphine) so the new drug is safer. **[1] (Marking of this question will depend on calculation in a). However, the examiner will use your answer in a) even if it is incorrect.)**

Page 41
QUICK TEST

1. Fungal disease
2. By cutting back or removing diseased trees
3. Fungicides or nitrogen fertilisers
4. Sap and leaves

EXAM PRACTICE

1. **Three from:** Make up 'acid rain' solution (e.g. a dilute sulfuric or nitric acid solution, or a solution with a pH in the range 2-6); Minimum of 2 groups of infected roses – spray one group with acid solution; Control variable e.g. same place in garden; Measurement/assessment technique e.g. number of bushes with blackspot infection [3]
2. Mechanical: **One from:** Thorns/hairs; Leaves that droop/curl on contact; Mimicry [1]
   Physical: **One from:** Tough, waxy leaves; Cellulose cell walls; Layers of dead cells on stems [1]

Page 43
QUICK TEST

1. 37°C
2. **Two from:** Osmoregulation/water balance; Balancing blood sugar levels; Maintaining a constant body temperature; Controlling metabolic rate.

EXAM PRACTICE

1. Pancreas – Produces insulin
   Skin receptor – Detects pressure
   Pituitary gland – Releases TSH
   Hypothalamus – Releases TRH
   [3 marks for 4 correct, 2 marks for 3 correct, 1 mark for 2 correct]

Page 45
QUICK TEST

1. Synapses
2. They ensure a rapid response to a threatening/harmful stimulus, e.g. picking up a hot plate. As a result, they reduce harm to the human body.

EXAM PRACTICE

1. a) X is the stimulus [1]
   b) It ensures a rapid response to a threatening/harmful stimulus [1] It is an unconscious reaction [1]
   c) **Four from:** Pain/pressure receptor in skin stimulated; Sensory neurone sends impulse; To spine; Intermediate neurone relays impulse to motor neurone; Motor neurone sends impulse to arm muscle; Arm muscle contracts [4]

Page 47
QUICK TEST

1. Thyroxine
2. Pancreas
3. Reduced ability of cells to absorb insulin and therefore high levels of blood glucose. This leads to tiredness, frequent urination, poor circulation, eye problems, etc.

EXAM PRACTICE

1. a) **Two from:** Causes all body cells to absorb glucose; Causes liver to convert glucose to glycogen; Blood glucose level is reduced [2]
   b) Insulin injections and monitoring of diet [1]
   c) The body's cells often no longer respond to insulin [1] It can be managed by adjusting carbohydrate intake [1]

Page 49
QUICK TEST

1. **Three from:** Glucose; Amino acids; Fatty acids; Glycerol; Some water
2. The pituitary gland

EXAM PRACTICE

1. a) Excretion [1]
   b) Amino acids [1]

Page 51
QUICK TEST

1. FSH, LH, progesterone, oestrogen
2. **One from:** Production of sperm in testes; Development of muscles and penis; Deepening of the voice; Growth of pubic, facial and body hair

EXAM PRACTICE

1. a) Oestrogen – promotes repair of uterus wall and stimulates egg release [1]
   Progesterone – maintains lining of uterus [1]
   b) Release of egg/ovulation [1]
2. Breast development [1] Sperm production [1] Menstrual cycle [1]

Page 53
QUICK TEST

1. Spermicidal agent. It is cheap.
2. They provide a barrier/prevent transferral of the virus during sexual intercourse.

EXAM PRACTICE

1. Advantage – very effective against STIs [1]
   Disadvantage – **One from:** Can only be used once; May interrupt sexual activity; Can break; May be allergic to latex [1]
2. Woman is given FSH and LH to stimulate the production of several eggs [1] Sperm and eggs are then introduced together outside the body in a petri dish [1] Successfully growing embryos can then be transplanted into the woman's uterus [1]

Answers

Page 55
QUICK TEST
1. Fertilisation
2. The species will become extinct.
3. **One from:** Produces clones of the parent – if these are successfully adapted individuals, then rapid colonisation and survival can be achieved; Only one parent required; Fewer resources required than sexual reproduction; Faster than sexual reproduction
4. When gametes join outside the body of the female

EXAM PRACTICE
1. **Advantage – One from:** Produces variation; Survival advantage when the environment is changing; Can be used by humans for selective breeding purposes [1]
   Disadvantage: One from: Slow process; Disadvantage in stable environments; More resources required; Results of selective breeding are unpredictable [1]
2. a) Not as dependent on one species/if one species is low in numbers or becomes extinct there is another organism available to act as a 'host'. [1]
   b) **One from:** Fungi; Mosses; Ferns; Protista [1]

Page 57
QUICK TEST
1. It has allowed the production of linkage maps that can be used for tracking inherited traits from generation to generation. This has led to targeted treatments for these conditions.
2. A, T, C and G

EXAM PRACTICE
1. By collecting and analysing DNA samples from different groups of peoples [1] human migration patterns can be deduced [1]
2. Backbone consisting of a sugar molecule [1] attached to a phosphate molecule [1] nitrogenous base attached [1]

Page 59
QUICK TEST
1. Alternative forms of a gene on a homologous pair of chromosomes
2. A dominant allele controls the development of a characteristic even if it is present on only one chromosome in a pair. A recessive allele controls the development of a characteristic only if a dominant allele is not present.
3. A, T, C, G

EXAM PRACTICE
1. A recessive allele controls the development of a characteristic only if a dominant allele is not present. [1] So, the recessive allele must be present on both chromosomes in a pair. [1]
2. Base sequence is altered [1] Amino acid sequence is also changed [1] Amino acid sequence affects protein shape and type. [1]

Page 61
QUICK TEST
1. 0%

EXAM PRACTICE
1.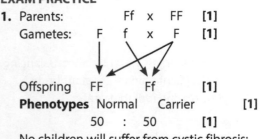
   Parents: Ff x FF   [1]
   Gametes: F   f   x   F   [1]

   Offspring   FF   Ff   [1]
   Phenotypes Normal   Carrier   [1]
                50 : 50   [1]
   No children will suffer from cystic fibrosis; 50% chance of a child being a carrier [1]

Page 63
QUICK TEST
1. **Two from:** The fossil record; Comparative anatomy (pentadactyl limb); Looking at changes in species during modern times; Studying embryos and their similarities; Comparing genomes of different organisms.
2. **Two from:** Catastrophic events, e.g. volcanic eruption; Changes to the environment; New predators; New diseases; New competitors

EXAM PRACTICE
1. Variation – horse-like ancestor adapted to environment/had different characteristics, named examples of different characteristics, [1] e.g. some horse-like mammals had more flipper-like limbs (as whales have flippers), mutation in genes allowed some individuals to develop these advantageous characteristics [1]
   Idea of competition for limited resources; examples of different types of competition e.g. obtaining food in increasingly water-logged environment [1]
   Idea of survival of the fittest – named examples of different adaptations, e.g. some horse-like mammals had more flipper-like limbs that allowed them to swim well in water [1]

Idea of successful characteristics being inherited – genes passed on to next generation, extinction – variety which is least successful does not survive and becomes extinct [1]

Page 65
QUICK TEST
1. The biological changes seen in species over millions of years
2. Ardi
3. **Any two examples**, e.g. Peppered moths; Antibiotic resistance in bacteria

EXAM PRACTICE
1. **Two from:** Changes in DNA/genes; Mutation; Production of new proteins; Proteins determine structural or behavioural adaptations [2]

Page 67
QUICK TEST
1. Genetic engineering is more precise and it takes less time to see results.
2. A plasmid is a ring of DNA found in bacteria.

EXAM PRACTICE
1. Advantage – **One from:** It results in an organism with the 'right' characteristics for a particular function; In farming and horticulture, it is a more efficient and economically viable process than natural selection [1]
Disadvantage – **One from:** Intensive selective breeding reduces the gene pool – the range of alleles in the population decreases so there is less variation; Lower variation reduces a species' ability to respond to environmental change; It can lead to an accumulation of harmful recessive characteristics (in-breeding) [1]
2. Stage 2: Gene removed using restriction enzyme [1]
Stage 3: Bacterial plasmid 'cut open' using restriction enzyme [1]
Stage 4: Ligase enzyme used to insert human gene into bacterial plasmid [1]
Stage 5: Bacterial cells containing plasmids rapidly reproduce [1]
Stage 6: Insulin purified from fermenter culture and produced in commercial quantities [1]

Page 69
QUICK TEST
1. Genus and species
2. Carl Woese

EXAM PRACTICE
1. Phylum [1] Order [1] Genus [1]
2. Canis familiaris [1]

Page 71
QUICK TEST
1. An ecosystem includes the habitat, its communities of animals and plants, together with the physical factors that influence them.
2. Food, mates, territory
3. **Any two suitable examples**, e.g. bacteria living in deep sea vents and icefish (exist in waters less than 0°C in temperature).

EXAM PRACTICE
1. i) Allow it to grip/stick to trees [1]
   ii) Camouflage so it won't be seen [1]
   iii) To catch insects/prey [1]
2. Extremophiles are organisms that can survive in extreme environmental conditions [1]
One from: Bacteria living in deep sea vents have optimum temperatures for enzymes that are much higher than 37°C; Icefish have antifreeze chemicals in their bodies which lower the freezing point of body fluids; Some organisms resist high salt concentrations or pressure [1]

Page 73
QUICK TEST
1. It would tell you how numbers of different species vary along the line of the transect.
2. Biotic factors are living, abiotic factors are non-living.

EXAM PRACTICE
1. a) To ensure they obtained representative samples [1]
   b) Working – numbers arranged in order of magnitude: 1,2,2,3,4,5,5,6
   The middle 2 numbers are 3 and 4, therefore the median is 3.5 **[3 marks for correct answer. If incorrect, 1 mark for each correct working]**
   c) Area of the field is 60 x 90 = 5400 m^2
   Therefore, number of daisies =
   5400 × 2.3 = 12,420
   [2 marks for correct answer. If incorrect, 1 mark for correct working]

Page 75
QUICK TEST
1. An animal that kills and eats other animals
2. The level that an organism feeds at, e.g. producer, primary consumer level
3. When green plants absorb it for photosynthesis
4. The green plant

Answers

EXAM PRACTICE

1. **a)** **One from:**

   Reed $\rightarrow$ Reed weevil $\rightarrow$ Spider

   Reed $\rightarrow$ Water beetle $\rightarrow$ Spider

   Flag iris $\rightarrow$ Water beetle $\rightarrow$ Spider

   Flag iris $\rightarrow$ Water cricket $\rightarrow$ Dragonfly **[1]**

   b) **One from:**

   Reed $\rightarrow$ Reed weevil $\rightarrow$ Spider $\rightarrow$ Songbird

   Reed $\rightarrow$ Water beetle $\rightarrow$ Spider $\rightarrow$ Songbird

   Flag iris $\rightarrow$ Water beetle $\rightarrow$ Spider $\rightarrow$ Songbird

   Flag iris $\rightarrow$ Water cricket $\rightarrow$ Dragonfly $\rightarrow$ Songbird

   [1]

2. As the rabbit population decreases, there is less food available for the stoats **[1]** so the stoat population decreases **[1]** fewer rabbits are eaten, so the rabbit population increases **[1]**

Page 77
QUICK TEST

1. Cutting down of trees/forest/woodland
2. **One from:** Sulfur dioxide; Nitrogen dioxide
3. Peat bogs

EXAM PRACTICE

1. **a)** **Three from:** Certain plant species/crops become scarce; Due to lack of pollinators; Other species which depend on these plants will be endangered/reduced in number; Biodiversity falls **[3]**

   b) **Two from:** Reduce usage of pesticides such as *neonicotinoids*; Plant more species which attract bumble bees (flowers etc.); Farm bees in hives and release into the wild; Disease prevention measures to act against foul-brood disease/isolate colonies/remove diseased colonies **[2]**

Page 79
QUICK TEST

1. Condensation and evaporation
2. 50%
3. To limit the decline of fish stocks

EXAM PRACTICE

1. **a)** Photosynthesis **[1]**

   b) Deomposition/decay **[1]**

   ii) Decomposition involves aerobic respiration **[1]** Carbon dioxide is a product of respiration **[1]**

2. Advantage – **One from:** Quotas limit the amount of fish caught allowing time for more fish to reproduce; Larger mesh sizes allow smaller fish to escape and reach reproductive age **[1]**
   Disadvantage – **One from:** Quotas are difficult to enforce; Quotas need to be applied for longer than at present **[1]**

Chemistry

Page 81
QUICK TEST

1. Atoms
2. A compound
3. Crystallisation

EXAM PRACTICE

1. **a)** $2Mg_{(s)} + O_{2\,(g)} \rightarrow 2MgO_{(s)}$

   Correctly balanced equation [1] Correct state symbols [1]

   b) Two or more elements (or compounds) that are together but not chemically combined. **[1]**

   c) Add water to the mixture and stir. **[1]** Filter to remove the insoluble magnesium oxide. **[1]** Heat/leave the magnesium chloride solution somewhere warm to allow the water to evaporate. **[1]**

Page 83
QUICK TEST

1. +1 / positive
2. 12 protons, 12 electrons and 12 neutrons
3. Isotopes

EXAM PRACTICE

1. In the plum pudding model, the electrons are embedded throughout a ball of positive charge. **[1]** In the nuclear model, there is a central nucleus containing the positive charge **[1]** and the electrons are located in energy levels/shells. **[1]**

Page 85
QUICK TEST

1. 2, 8, 8, 1
2. He also predicted that there were more elements to be discovered. He predicted their properties and left appropriate places in the periodic table based on his predictions.

EXAM PRACTICE

1. **a)** 2,8,2 **[1]**

   b) After. **[1]** Element X has 12 electrons (and therefore 12 protons). The periodic table is arranged by increasing atomic number and 14 is greater than 12. **[1]**

Page 87
QUICK TEST

1. **Examples:** Airships, balloons, light bulbs, lasers, advertising signs
2. Potassium hydroxide and hydrogen

3. The reactivity decreases

EXAM PRACTICE

1. a) Three from: Sodium floats; Moves about the surface; Melts; Fizzes/effervescence **[3]**

  b) $2Na_{(s)} + 2H_2O_{(l)} \rightarrow 2NaOH_{(aq)} + H_{2\,(g)}$
Correct formulae [1] Correct state symbols [1]

2. a) Bromine is formed **[1]** which is brown. **[1]**

  b) Chlorine is more reactive than bromine **[1]** and so will take the place of/displace bromine in a compound. **[1]**

Page 89
QUICK TEST

1. Covalent

2. Two

3. A free-moving electron

4. The electrostatic force of attraction between two oppositely charged ions

EXAM PRACTICE

1. a) Potassium atoms lose their outer shell electron and sulfur atoms gain two electrons **[1]** forming K^+ ions that have the electronic configuration 2,8,8 **[1]** and S^{2-} ions that have the electronic configuration 2,8,8 **[1]**

  b) K_2S **[1]**

2. a)

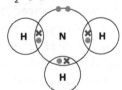

3 atoms of hydrogen each with two electrons. **[1]** Pair of electrons in the outer shell of nitrogen. **[1]**

  b) Atoms of neon have full outer shells of electrons **[1]** and so don't need to gain, lose or share any electrons. **[1]**

3. A lattice of metal cations **[1]** surrounded by delocalised electrons. **[1]**

Page 91
QUICK TEST

1. Any compound containing a metal and a non-metal, e.g. sodium chloride.

2. Electrostatic forces/strong forces between cations and anions

3. Simple molecular, polymers, giant covalent

EXAM PRACTICE

1. a) XCl_2 **[1]** There are twice as many chloride ions as there are X ions. **[1]**

  b) One from: Only a few ions are shown; The ions are not connected by 'lines'/chemical bonds are not solid objects; The ions are not held apart; Diagram doesn't accurately show relative size of particles. **[1]**

2. a) Covalent **[1]**

  b) Simple molecular **[1]**

  c) Giant covalent **[1]**

Page 93
QUICK TEST

1. solid, liquid, gas

2. The intermolecular forces/forces between the molecules

EXAM PRACTICE

1. a) -40°C **[1]**

  b) Liquid **[1]**

  c) Condensing/condensation **[1]**

2. In a solid the ions are in a fixed position/unable to move. **[1]** In the liquid state the ions are free to move **[1]** and carry the charge. **[1]**

Page 95
QUICK TEST

1. Because there are strong forces of attraction between the metal cations and delocalised electrons.

2. Two from: Drug delivery into the body; Lubricants; Reinforcing materials

EXAM PRACTICE

1. a) A mixture of two or more metals, or a mixture of a metal and a non-metal. **[1]**

  b) The layers of atoms are not able to slide over each other easily **[1]** because the different atoms have different sizes **[1]** and so the layers aren't regular.

2. a) Layers of carbon atoms **[1]** that each form three covalent bonds to other carbon atoms **[1]** There are weak intermolecular forces between the layers. **[1]**

  b) Each carbon atom only forms three bonds/has a spare/delocalised electron **[1]** which can move and carry the charge. **[1]**

Page 97
QUICK TEST

1. The same

2. The sum of the atomic masses of the atoms in a formula

3. Lower

4. 58

EXAM PRACTICE

1. a) 84 **[1]**

  b) (10-4.76) = 5.24g **[1]**

Answers

c) Thermal decomposition or endothermic [1]

d) It will increase [1] because oxygen is being added to the metal. [1]

Page 99
QUICK TEST
1. Li_2O
2. P_4O_{10}

EXAM PRACTICE
1. Ratio method:

	4Al	3O$_2$	2Al$_2$O$_3$
Number of moles reacting	4	3	2
Relative formula mass	27	32	102
Mass reacting/ formed (g)	108 ↘ ÷25	96	204 ↘ ÷25
Reacting mass	4.32 ↙		8.16 ↙

Correct answer [3]

Moles method:

Number of moles of aluminium reacting

$= \dfrac{4.32}{27} = 0.16$ [1]

Due to 4:2 ratio the number of moles of aluminium oxide formed = 0.08 [1]

Therefore, the mass of aluminium oxide formed = $0.08 \times 102 = 8.16$g [1]

Page 101
QUICK TEST
1. 0.84 g
2. $Si + 2Cl_2 \rightarrow SiCl_4$
3. A solid that dissolves in a liquid to form a solution

EXAM PRACTICE
1. Solution A has a concentration of

$\dfrac{27}{820} \times 1000 = 33$ g/dm^3 [1]

Solution B has a concentration of

$\dfrac{19}{560} \times 1000 = 34$ g/dm^3 [1]

Therefore, solution B has the highest concentration in g/dm^3 [1]

2.

Chemical	Fe	O$_2$	X
M$_r$	56	32	232
Moles = $\dfrac{mass}{M_r}$	$\dfrac{16.8}{56} = 0.3$	$\dfrac{6.4}{32} = 0.2$	$\dfrac{23.2}{232} = 0.1$ [1]
÷0.1	$\dfrac{0.3}{0.1} = 3$	$\dfrac{0.2}{0.1} = 2$	$\dfrac{0.1}{0.1} = 1$ [1]

Therefore, the balanced equation is

$3Fe + 2O_2 \rightarrow X$ [1]

Page 103
QUICK TEST
1. **Two from**: Zinc; Iron; Copper
2. Magnesium
3. aluminium + oxygen → aluminium oxide
4. Sodium (because it is more reactive)

EXAM PRACTICE
1. a) Magnesium is more reactive than copper, so is able to displace copper from compounds [1]

b) Copper(II) oxide [1] as it loses oxygen. [1]
[**Allow** copper/Cu^{2+} [1] as it gains electrons [1]]

Page 105
QUICK TEST
1. A method used to separate a soluble solid from its solution when you want to collect the solid.
2. Fe
3. aluminium + sulfuric acid → aluminium sulfate + hydrogen
4. Zinc nitrate
5. Calcium sulfate

EXAM PRACTICE
1. a) $MgO_{(s)} + 2HCl_{(aq)} \rightarrow MgCl_{2\,(aq)} + H_2O_{(l)}$
Correct formulae [1] Correctly balanced [1]

b) **Five from:** Using a measuring cylinder, transfer a known volume of hydrochloric acid to a beaker; Warm the acid; Add a spatula of magnesium oxide powder and stir until it dissolves; Repeat until no more magnesium oxide dissolves; Filter the mixture to remove excess magnesium oxide; Heat the filtrate/leave it somewhere warm to remove the water/allow crystallisation to occur. [5]

Page 107
QUICK TEST
1. A liquid or solution containing ions that is broken down during electrolysis.
2. H^+
3. Alkaline/alkali
4. A strong acid fully ionises/dissociates in solution. A weak acid only partially ionises/dissociates in solution.
5. Cations

EXAM PRACTICE
1. a) Pink [1] to colourless [1]

b) The hydroxide ion/OH$^-$ [1]

c) $H^+_{(aq)} + OH^-_{(aq)} \rightarrow H_2O_{(l)}$ [1]

d) Strong acids fully/completely [1] dissociate/ionise [1] in solution.

Page 109

QUICK TEST

1. Copper will be formed at the cathode; chlorine will be formed at the anode.
2. Hydrogen/H^+ ions and hydroxide/OH^- ions

EXAM PRACTICE

1. Product at cathode = copper [1]
   Product at anode = oxygen [1]

Page 111

QUICK TEST

1. Exothermic – **One from:** Combustion; Neutralisation; Oxidation; Precipitation; Displacement
   Endothermic – **One from:** Thermal decomposition; Citric acid; Sodium hydrogen carbonate
2. Given out

EXAM PRACTICE

1.
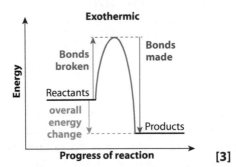

[3]

2. Bonds broken:
   H–H   436
   Cl–Cl   239
   Total = 675 [1]

   Bonds formed:
   2 H—Cl = 2 x 427 = 854 [1]
   Energy change = 675-854 = -179 kJ/mol [1]

Page 113

QUICK TEST

1. 3 g/s
2. **Two from**: The concentrations of the reactants in solution; The pressure of reacting gases; The surface area of any solid reactants; Temperature; Presence of a catalyst

EXAM PRACTICE

1. a) $\frac{90}{42}$ [1] = 2.1 [1] cm³/s
   Answer must be to 1 decimal place.
   b) It would decrease/halve. [1]

c) **One from:** Increasing the surface area of the magnesium/using magnesium powder; Increasing the temperature; Adding a catalyst [1]

Page 115

QUICK TEST

1. The minimum amount of energy that the particles must have when they collide in order to react.
2. There are fewer particles in the same volume of liquid and so there are fewer chances of reactant particles colliding.
3. They provide an alternative pathway of lower activation energy.

EXAM PRACTICE

1. a) The particles/molecules have more kinetic energy/move more quickly [1] meaning there will be more collisions per unit of time. [1] At a higher temperature more of the particles will possess energy equal to or greater than the activation energy for the reaction meaning there will be more successful collisions. [1]
   b) It provides an alternative reaction pathway [1] of lower activation energy. [1]
   c) At higher pressure there will be more molecules of gas per unit of volume [1] meaning more collisions per unit of time. [1]

Page 117

QUICK TEST

1. Endothermic
2. Equilibrium moves to the right-hand side / favours the forward reaction.

EXAM PRACTICE

1. $N_2 + 3H_2 \rightleftharpoons 2NH_3$
   Correct balanced equation [1] Use of $\rightleftharpoons$ symbol [1]
   b) The rate [1] of the forward reaction is equal to the rate of the reverse reaction. [1]
2. a) Increasing pressure favours the reaction/moves the equilibrium position to the side with the fewer moles of gas. [1] Therefore, the equilibrium position will move to the left-hand side/the reverse reaction will be favoured. [1]
   b) Increasing the temperature favours the endothermic reaction. [1] In this case the forward reaction is endothermic (indicated by a positive ΔH value) and so increasing the temperature will move the position of equilibrium to the right-hand side/the forward reaction will be favoured. [1]

Answers

QUICK TEST
1. Dead biomass
2. C_nH_{2n+2}.
3.

H   H
| |
H – C – C – H
| |
H   H

EXAM PRACTICE
1. **a)** To separate it into useful fractions/components **[1]**
    b) The crude oil is heated until it boils/vaporises. **[1]** It then enters the fractionating column where there is a temperature gradient. **[1]** Molecules condense at different levels in the fractionating column according to their boiling point. **[1]**

Page 121
QUICK TEST
1. Using a catalyst
2. Fuels

EXAM PRACTICE
1. **a)** $C_8H_{18} \rightarrow C_4H_{10} + 2C_2H_4$
    Correct formulae **[1]** Balanced equation **[1]**
    b) The alkene will decolourise **[1]** bromine water. **[1]**

Page 123
QUICK TEST
1. To separate mixtures of dyes
2. A mixture that has been designed as a useful product
3. **Two from:** Fuels; Cleaning materials; Paints; Medicines; Foods; Fertilisers
4. 0.7

EXAM PRACTICE
1. **a)** Green and yellow **[1]**
    b) By mixing **[1]** purple and yellow **[1]** inks.

Page 125
QUICK TEST
1. Oxygen
2. Add a lighted splint and there will be a squeaky pop.
3. Limewater
4. It turns milky/cloudy.

EXAM PRACTICE
1. Test the gas with damp **[1]** Litmus paper. **[1]** It will be bleached if chlorine is present. **[1]**

Page 127
QUICK TEST
1. As a product of photosynthesis
2. **Two from:** Carbon dioxide; Water vapour; Methane; Ammonia; Nitrogen, Sulfur dioxide
3. Approximately 0–3%
4. Carbon dioxide

EXAM PRACTICE
1. Due to photosynthesis. **[1]** The amount of carbon dioxide has decreased. **[1]** The amount of oxygen has increased. **[1]**
2. **a)** 16 cm^3 **[1]**
    b) **Two from:** Carbon dioxide; Water (vapour); Argon; Any other named noble gas **[2]**
    c) $\frac{164}{820} \times 100 = 20\%$ **[1]**

Page 129
QUICK TEST
1. **Two from:** Water vapour; Carbon dioxide; Methane
2. **Two from:** Rising sea levels leading to flooding/coastal erosion; More frequent/severe storms; Changes to the amount, timing and distribution of rainfall; Temperature and water stress for humans and wildlife; Changes in the food producing capacity of some regions; Changes to the distribution of wildlife species
3. The total amount of carbon dioxide (and other greenhouse gases) emitted over the full life cycle of a product, service or event.

EXAM PRACTICE
1. **Three from:** Combustion of fossil fuels leads to greater carbon dioxide emissions into the atmosphere; Deforestation reduces the amount of carbon dioxide removed from the atmosphere by photosynthesis; Increased animal farming releases more methane into the atmosphere; Increased use of landfill sites releases more methane into the atmosphere **[3]**

Page 131

QUICK TEST

1. **Three from**: Carbon dioxide; Carbon monoxide; Water vapour; Sulfur dioxide; Nitrogen oxides
2. From the oxidation/combustion of sulfur in fuels

EXAM PRACTICE

1. a) Incomplete combustion of petrol/diesel/fuel [1]
   b) As the car gets older less oxygen is able to get to burn the fuel e.g. blocked pipes etc. [1]
   c) They can cause respiratory problems. [1]
   d) Global dimming [1] Lung damage [1]

Page 133

QUICK TEST

1. Living such that the needs of the current generation are met without compromising the ability of future generations to meet their own needs.
2. Pure water has no chemicals added to it. Potable water may have other substances in it but it is safe to drink.
3. Screening and grit removal; Sedimentation; Anaerobic digestion of sludge; Aerobic biological treatment of effluent.

EXAM PRACTICE

1. a) Distillation [1]
   b) A = sea water/salty water [1]
      B = condenser [1]
      C = pure/desalinated water [1]
   c) Reverse osmosis [1]
   d) They require lots of energy. [1]

Page 135

QUICK TEST

1. **One from**: Electrical wiring; Water pipes
2. Phytomining, bioleaching
3. An extraction method that uses bacteria to extract metals from low-grade ores.
4. **One from**: Displacement using scrap iron; Electrolysis

EXAM PRACTICE

1. a) Supplies of copper rich ores are diminishing. [1] Bioleaching is a method of obtaining copper from low-grade ores. [1]
   b) By displacement/adding a more reactive metal e.g. (scrap) iron. [1] By electrolysis/passing electricity through the solution. [1]
   c) Phytomining [1]

Page 137

QUICK TEST

1. The environmental impact of a product over the whole of its life.
2. **Two from**: How much energy is needed; How much water is used; What resources are required; How much waste is produced; How much pollution is produced
3. An industrial method of extracting iron from iron ore
4. **One from**: Using fewer items that come from the earth; Reusing items; Recycling more of what we use

EXAM PRACTICE

1. **Three from:** The production of carbon dioxide is only one measure in a LCA; Other pollutants may be produced in the production of plastic bags; No reference is made of energy requirements; No reference is made to water usage; Plastic bags may be used many more times than paper bags/plastic bags have a longer life span than paper bags; The LCA is incomplete; More trees may be grown to produce more paper for bags which absorb CO_2 [3]

Answers

Physics

QUICK TEST
1. A vector quantity has both a magnitude and a direction.
2. A scalar quantity only has a magnitude.
3. $0.033 \times 10 = 0.33$ N
4. 57 J

EXAM PRACTICE
1. a) Mass = weight / gravitational field strength
   $$= 560 / 9.8 \text{ [1]}$$
   $$= 57.1 \text{kg [1]}$$
   b) Work done = force × distance moved
   $$= 560 \times 4 \text{ [1]}$$
   $$= 2240 \text{J [1]}$$
   c) Pushing the crate is a contact force as two objects are physically touching [1] whilst magnetic force does not require objects to be physically touching. [1]

Page 141
QUICK TEST
1. A stretched object that returns to its original length after the force is removed
2. $0.3 \times 2 = 0.6$ N
3. Elastic potential energy

EXAM PRACTICE
1. a) Extension = force / spring constant
   $$= 7 / 18 \text{ [1]}$$
   $$= 0.39 \text{m [1]}$$
   b) i) A straight line [1]
   ii) A linear relationship [1]
   iii) A non-linear relationship [1] as all the points could no longer be connected by a straight line [1]

Page 143
QUICK TEST
1. Speed has a magnitude but not a direction.
2. $\frac{15}{3} = 5$ km/h
3. 330 m/s

EXAM PRACTICE
1. a) Distance = speed × time
   $$= 21 \times 56 \text{ [1]}$$
   $$= 1176 \text{m [1]}$$
   b) Time = distance / speed
   $$= 14000 / 15 \text{ [1]}$$
   $$= 933 \text{ seconds [1]}$$

$933 / 60 = 15.6$ minutes
   Time = 15.6 minutes [1]
   c) Her velocity changes [1] as although her speed remains constant the direction she is travelling in is changing. [1]

Page 145
QUICK TEST
1. Change in velocity over time
2. A negative acceleration/deceleration
3. Speed

EXAM PRACTICE
1. a) Change in velocity = 54-0
   $$= 54 \text{m/s [1]}$$
   Average acceleration = $54 / 16 = 3.4 \text{m/s}^2$ [1]
   b) They have reached terminal velocity [1] so the resultant force is 0. [1]

Page 147
QUICK TEST
1. $93 \times 10 = 930$ N
2. 83 N
3. It continues to move at the same speed.
4. It remains stationary.

EXAM PRACTICE
1. a) Acceleration = force / mass
   $$= 2100 / 7 \text{ [1]}$$
   $$= 300 \text{ m/s}^2 \text{ [1]}$$
   b) 2100N [1] as Newton's third law states [1] that when two objects interact they exert equal and opposite forces on each other. [1]

Page 149
QUICK TEST
1. Thinking distance and braking distance
2. **Two from**: Rain; Ice; Snow
3. It may lead to brakes overheating and/or loss of control.

EXAM PRACTICE
1. a) Momentum = mass × velocity
   Momentum = 1080×7.5 [1]
   Momentum = 8100 kg m/s [1]
   b) Mass = momentum / velocity
   Mass = 9675 / 7.5 [1]
   Mass = 1290kg [1]
   c) The total momentum of both cars is 9675 kg m/s. [1] As the momentum of the moving car was 9675 kg m/s and the momentum of the stationary car was 0. [1] Momentum is conserved so the total momentum before the event is the total momentum after the event. [1]

Page 151
QUICK TEST
1. It is the amount of energy required to raise the temperature of one kilogram of a substance by one degree Celsius.
2. $0.5 \times 0.62 \times 5^2 = 7.75$ J
3. $17 \times 456 \times 10 = 77.52$ kJ

EXAM PRACTICE
1. a) Elastic potential energy = 0.5 × spring constant × extension^2
$$= 0.5 \times 115 \times 0.49^2 \text{ [1]}$$
$$= 13.8N \text{ [1]}$$
   b) Elastic potential to kinetic energy [1] and gravitational potential energy. [1]
   c) 0 [1] as the ball is no longer raised above the ground. [1]

Page 153
QUICK TEST
1. Material used to reduce transfer of heat energy
2. Power = work done/time
$$= 50/2$$
$$= 25 \text{ W}$$
3. It reduces the thermal conductivity.

EXAM PRACTICE
1. $\frac{600}{802} = 0.75 = 75\%$ [1]

Page 155
QUICK TEST
1. **Three from**: Biofuel; Wind; Hydro-electricity; Geothermal; The tides (tidal power); The sun (solar power); Water waves
2. **Three from**: Coal; Oil; Gas; Nuclear fuel
3. The wind doesn't always blow and it's not always sunny, so electricity isn't always generated.
4. They produce carbon dioxide, which is a greenhouse gas. Increased greenhouse gas emissions are leading to climate change. Particulates and other pollutants are also released, which cause respiratory problems.

EXAM PRACTICE
1. a) Renewable energy resources will not run out [1] and do not produce carbon dioxide, which is a greenhouse gas that can lead to climate change. [1] Nuclear power plants also do not produce carbon dioxide [1] and are very reliable so can generate electricity even if renewable energy resources are not, e.g. if the wind isn't blowing. [1]
   b) Nuclear power plants produce hazardous nuclear waste [1] and have the potential for devastating nuclear accidents. [1]
   c) Some people consider wind farms to be ugly and to spoil the landscape [1] building tidal power stations can destroy important tidal habitats. [1]

Page 157
QUICK TEST
1. **One from**: Sound wave; Wave in a stretched spring
2. **One from**: Water wave; Electromagnetic wave
3. The distance from a point on one wave to the equivalent point on an adjacent wave.
4. Amplitude, wavelength, frequency, period

EXAM PRACTICE
1. a) Water waves [1]
   b) period = 1 / frequency
$$= 1 / 4 \text{ [1]}$$
$$= 0.25 \text{ seconds [1]}$$
   c) wave speed = frequency × wavelength
$$= 4 \times 0.35 \text{ [1]}$$
$$= 1.4 \text{ m/s [1]}$$
   d) This shows that waves transfer energy and information [1] without transferring matter. [1]

Page 159
QUICK TEST
1. Infrared
2. Visible light
3. Transverse
4. The change in direction of a wave as it travels from one medium to another
5. Towards the normal

EXAM PRACTICE
1. Visible light and UV radiation are both examples of electromagnetic waves. [1] All electromagnetic waves travel at the same speed through the vacuum of space. [1]
2. Light refracts as it moves between the water and the air. [1] As light travels faster in air than water the light bends away from the normal as it moves from water to the air. [1] This means the fish are not exactly where they appear to be. [1]

Page 161
QUICK TEST
1. 1000 millisieverts = 1 sievert
2. By oscillations in electrical circuits
3. Sterilising, medical imaging, treating cancer
4. **One from**: Electrical heaters; Cooking food; Infrared cameras; Television remote controls

Answers

EXAM PRACTICE

1. X-rays are absorbed differently by different parts of the body. **[1]** More are absorbed by hard tissues e.g. bones **[1]** and less are absorbed by soft tissues. This allows an image of the inside of the body to be created. **[1]**

2. Ultraviolet radiation can lead to premature ageing of skin **[1]** and increased risk of skin cancer. **[1]**

Page 163
QUICK TEST

1. ─▭─
2. 3.8 C
3. 600/3 = 200 A

EXAM PRACTICE

1. a) A battery or a cell **[1]** as this is a source of energy which produces a potential difference. **[1]**

   b) i) Time = charge flow / current
      = 160 / 8 **[1]**
      = 20 seconds **[1]**

   ii) 8A **[1]** as the current is the same at any point in a closed circuit. **[1]**

Page 165
QUICK TEST

1. Because resistance depends on light intensity
2. 6 × 3 = 18 V
3. 8/2 = 4 Ω

EXAM PRACTICE

1. a) An ammeter **[1]** and a voltmeter **[1]**

   b) Current = potential difference / resistance
      = 12 / 4 **[1]**
      = 3A **[1]**

   c) The temperature of the filament increased. **[1]** This would cause the resistance of the lamp to increase. **[1]**

Page 167
QUICK TEST

1. Add the resistances of both resistors together.
2. 4 A
3. A circuit that contains branches and where all the components have the same voltage.

EXAM PRACTICE

1. a) 9V **[1]** as all the components in a parallel circuit have the same potential difference across them. **[1]**

   b) The total resistance of all the components is less than 3 ohms **[1]** because in a parallel circuit the total resistance of all the components is less than the resistance of the smallest individual resistor. **[1]**

Page 169
QUICK TEST

1. Alternating current and direct current
2. Circuit breakers can be reset and operate faster than a fuse.
3. Brown
4. 0V

EXAM PRACTICE

1. a) No **[1]** The appliance was double insulated **[1]** which means it is impossible for the case to become live as it's plastic. **[1]** The appliance therefore did not require an earth wire. **[1]**

   b) In the case of a fault a large current flows from the live wire to the earth wire. **[1]** This melts the fuse and disconnects the live wire. **[1]**

Page 171
QUICK TEST

1. They lower the potential difference of the transmission cables to a safe level for domestic use.
2. 50 × 10 = 500 J
3. $4^2 \times 3 = 48$ W

EXAM PRACTICE

1. a) Resistance = power / current^2
      = 8000000 / 500^2 **[1]**
      = 32 ohms **[1]**

   b) Energy transferred = power × time
      = 8000000 × 120 **[1]**
      = 960000 kJ **[1]**

   c) The current is kept relatively low to reduce energy loss due to heating in the cables. **[1]** This increases the efficiency of the energy transmission. **[1]**

Page 173
QUICK TEST

1. They repel
2. They attract
3. A non-contact force
4. The Earth's core is magnetic and produces a magnetic field.

EXAM PRACTICE

1. **a)** The region around a magnet where a force acts on another magnet or on magnetic material. **[1]**
   b) At the poles of the magnet **[1]**
   c) A force of attraction **[1]**
   d) From the north pole to the south pole **[1]**

Page 175
QUICK TEST

1. The current through the wire and the distance from the wire
2. Magnetic field, force and current

EXAM PRACTICE

1. Force = magnetic flux density × current × length
   = 3.2 × 8 × 9 **[1]**
   = 230.4N **[1]**

Page 177
QUICK TEST

1. Solid, liquid, gas
2. $\frac{4}{0.002}$ = 2000 kg/m^3
3. The mass stays the same.
4. True

EXAM PRACTICE

1. **a)** As the temperature of the gas increased, the pressure of the gas also increased. **[1]**
   b) The volume of the gas should be kept constant. **[1]**
   c) The gas would condense to form a liquid. **[1]**
   d) Cooling the gas was an example of change of state. **[1]** If this change is reversed, a substance recovers its original properties. **[1]**

Page 179
QUICK TEST

1. 300 × 126 = 37 000 kJ
2. 301 K
3. 17°C

EXAM PRACTICE

1. **a)** The internal energy of the system increases. **[1]**
   b) The internal energy of the system is equal to the total kinetic energy **[1]** and potential energy **[1]** of all the atoms and molecules that make up the system.
   c) The substance was changing state so there was no change in temperature. **[1]**

Page 181
QUICK TEST

1. Protons – positive; Neutrons – no charge; Electrons – negative
2. Isotopes
3. False. Atoms turn into positive ions if they lose outer electrons.

EXAM PRACTICE

1. **a)**

Particle	Relative mass
Proton	1
Neutron	1
Electron	0.0005

[All correct 2 marks, 1 mark if 1 incorrect]
   b) The nucleus is positively charged **[1]** whilst the atom has no overall charge. **[1]**
   c) They become excited and move to a higher energy level **[1]** further from the nucleus. **[1]**

Page 183
QUICK TEST

1. A few metres
2. A thin sheet of paper
3. The mass of the nucleus doesn't change but the charge of the nucleus does change.

EXAM PRACTICE

1. **a)** Alpha decay **[1]**
   b) A = 64 **[1]** B = 145 **[1]**
   c) Alpha decay causes the mass and charge of the nucleus to decrease. **[1]** Gamma ray emission does not cause the mass or charge of the nucleus to change. **[1]**

Page 185
QUICK TEST

1. Exposing an object to nuclear radiation without the object becoming radioactive itself.
2. The radioactive atoms decay and release radiation.

EXAM PRACTICE

1. **a)** 3178 / 2 = 1589 **[2]**
   b) 3178 / 2 = 1589 = count after 6 hours
      1589 / 2 = 794.5 **[1]** count after 12 hours
      794.5 / 2 = 397.25 count after 18 hours
      397.25 = 198.625 **[1]** count after 24 hours
   c) i) This is contamination **[1]** as it's the unwanted presence of materials containing radioactive atoms. **[1]**
      ii) So the results can be checked by other scientists **[1]** in the peer review process **[1]**

Glossary

Biology

Active ingredient – Chemical in a drug that has a therapeutic effect (other chemicals in the drug simply enhance flavour or act as bulking agents)

Active site – The place on an enzyme molecule into which a substrate molecule fits

Alleles – Alternative forms of a gene on a homologous pair of chromosomes

Anatomy – The study of structures within the bodies of organisms

Anthropologists – Scientists who study the human race and its evolution

Antibodies – Proteins produced by white blood cells (particularly lymphocytes). They lock on to antigens and neutralise them

Antigen – Molecular marker on a pathogen cell membrane that acts as a recognition point for antibodies

Aphid – A type of sap-sucking insect

Binomial system – 'Two name' system of naming species using Latin

Biomass – Mass of organisms calculated by multiplying their individual mass by the number that exist

Biosphere – Area on the Earth's crust that is inhabited by living things

Carbon sinks – Resources that lock up carbon in their structure rather than allowing them to form carbon dioxide, e.g. peat bogs, oceans, limestone deposits

Catalyst – A substance that controls the rate of a chemical reaction without being chemically changed itself

Cellulose – Large carbohydrate molecule found in all plants; an essential constituent of cell walls

Chlorophyll – A molecule that gives plants their green colour and absorbs light energy

Chlorosis – Where leaves lose their colour as a result of mineral deficiency

Cilia – Microscopic hairs found on the surface of epithelial cells; they 'waft' from side to side in a rhythmic manner

Clone – A genetically identical cell, tissue, organ or organism

Colony – A large number of microorganisms (of one type) growing in a location, e.g. a circular colony on the surface of agar

Common ancestor – Organism that gave rise to two different branches of organisms in an evolutionary tree

Competition – When two individuals or populations seek to exploit a resource, e.g. food. One individual/population will eventually replace the other

Compost – Fertiliser produced from the decay of organic plant material

Compress – Squash or squeeze. In geology, this is usually due to Earth movements or laying down sediments

Contraception – Literally means 'against conception'; any method that reduces the likelihood of a sperm meeting an egg

Coronary artery – The blood vessel delivering blood to the heart muscle

Daughter cells – Multiple cells arising from mitosis and meiosis

Decomposers – Microorganisms that break down dead plant and animal material by secreting enzymes into the environment. Small, soluble molecules can then be absorbed back into the microorganism by the process of diffusion

Differentiation – A process by which cells become specialised, i.e. have a particular structure to carry out their function

Diploid – A full set of chromosomes in a cell (twice the haploid number)

Droplet infection – Transmission of microorganisms through the aerosol (water droplets) produced through coughing and sneezing

Emissions – Gaseous products usually connected with pollution, e.g. carbon dioxide emissions from exhausts

Endangered – Category of risk attached to rare species of plants and animals. This usually triggers efforts to preserve the species' numbers

Endocrine system – System of ductless organs that release hormones

Endothermic – A change that requires the input of energy

Energy demand – Energy required by tissues (particularly muscle) to carry out their functions

Glossary

Environmental resources – Materials or factors that organisms need to survive, e.g. high oxygen concentration, living space or a particular food supply

Epidermis – Outer layer of tissue in the skin

Epithelial – A single layer of cells often found lining respiratory and digestive structures

Eutrophication – A process where nitrates and phosphates enrich waterways, causing massive growth of algae and loss of oxygen

(Evapo)transpiration – Evaporation of water from stomata in the leaf

Exothermic reaction – A reaction that gives out heat

External fertilisation – Gametes join outside the body of the female

Extinction – When there are no more members of a species left living

Fittest – The most adapted individual or species

Follicle stimulating hormone – A hormone produced by the pituitary gland that controls oestrogen production by the ovaries

Fusion – Joining together; in biology the term is used to describe fertilisation

Gamete – A sex cell, i.e. sperm or egg

Glucagon – Hormone released by the pancreas that stimulates the conversion of glycogen to glucose

Glycogen – Storage carbohydrate found in animals

Gravitropism (geotropism) – Growth response in plants against or with the force of gravity

Haemoglobin – Iron-containing molecule that binds to oxygen molecules in red blood cells

Haploid – A half set of chromosomes in a cell; haploid cells are either eggs or sperm

Haploid nucleus – Nucleus with a half chromosome set

Herbicide – A chemical applied to crops to kill weeds

Herbicide – Chemical sprayed on crops to kill weeds

Herd immunity – Vaccination of a significant portion of a population (or herd) makes it hard for a disease to spread because there are so few people left to infect. This gives protection to susceptible individuals such as the elderly or infants

Homologous chromosomes – A pair of chromosomes carrying alleles that code for the same characteristics

Hydrogen bond – A bond formed between hydrogen and oxygen atoms on different molecules close to each other

Immune system – A system of cells and other components that protect the body from pathogens and other foreign substances

Immunological memory – The system of cells and cell products whose function is to prevent and destroy microbial infection

Implantation – Process in which an embryo embeds itself in the uterine wall

Impulses – Electrical signals sent down neurones that trigger responses in the nervous system

Inhibition – The effect of one agent against another in order to slow down or stop activity, e.g. chemical reactions can be slowed down using inhibitors. Some hormones are inhibitors

Internal fertilisation – Gametes join inside the body of the female

Intrauterine – Inside the uterus

Kinetic energy – Energy possessed by moving objects, e.g. reactant molecules such as enzyme and substrate molecules

Lignin – Strengthening, waterproof material found in walls of xylem cells

Limiting factor – A variable that, if changed, will influence the rate of reaction most

Malnutrition – A diet lacking in one or more food groups

Menstruation – Loss of blood and muscle tissue from the uterus wall during the monthly cycle

Meristems – Growth regions in plants where stem cells are produced, e.g. apical bud

Mucus – Thick fluid produced in the lining of respiratory and digestive passages

Non-communicable – Disease or condition that cannot be spread from person to person via pathogen transfer

Oral contraceptive – Hormonal contraceptive taken in tablet form

Organelle – A membrane-bound structure within a cell that carries out a particular function

Oxygen debt – The oxygen needed to remove lactic acid after exercise

Partially permeable membrane – A membrane with microscopic holes that allows small particles through (e.g. water) but not large ones (e.g. sugar)

Pathogen – A harmful microorganism

Pesticide – Chemical sprayed on crops to kill invertebrate pests

Pharmaceutical drug – Chemicals that are developed artificially and taken by a patient to relieve symptoms of a disease or treat a condition

Photosynthetic organism – Able to absorb light energy and manufacture carbohydrate from carbon dioxide and water

Placebo – A substitute for the medication that does not contain the active ingredient

Plaque – Fatty deposits that can build up in arteries

Plasmid – A ring of DNA found in bacteria

Polymer – A long chain molecule made up of individual units called monomers

Receptor molecule – Protein on the outer membrane of a cell that binds to a specific molecule, such as transmitter substance

Resolution – The smallest distance between two points on a specimen that can still be told apart

Sample – A small area or population that is measured to give an indication of numbers within a larger area or population

Spinal cord – Nervous tissue running down the centre of the vertebral column; millions of nerves branch out from it

Stem cells – Undifferentiated cells that can become any type of cell

Stimuli – Changes in the internal or external environment that affect receptors

Substrate – The molecule acted on by an enzyme

Surface area to volume ratio – A number calculated by dividing the total surface area of an object by its volume. When the ratio is high, the efficiency of diffusion and other processes is greater

Sustainability – Carrying out human activity, e.g. farming, fishing and extraction of resources from the ground, so that damage to the environment is minimised or removed

Symptoms – Physical or mental features that indicate a condition or disease, e.g. rash, high temperature, vomiting

Therapeutic cloning – A process where embryos are grown to produce cells with the same genes as a particular patient

Transpiration – Flow of water through the plant ending in evaporation from leaves

Turgidity – Where plant cells fill with water and swell as a result of osmosis

Urea – Nitrogenous waste product

Vectors – Small organisms (such as mosquitoes or ticks) that pass on pathogens between people or places

Ventilation – Process of drawing air into and out of the lungs. It involves the ribs, intercostal muscles and diaphragm

Vertebrate – Animal with a backbone

Water potential/diffusion gradient – A higher concentration of particle numbers in one area than another; in living systems, these areas are often separated by a membrane or cell wall

Yield – The weight of living material harvested in farming and fishing

Chemistry

Activation energy – The minimum amount of energy that particles must collide with in order to react

Alloy – A mixture of two or more metals, or a mixture of a metal and a non-metal

Anion – A negative ion

Anode – The positive electrode

Atom – The smallest part of an element that can enter into a chemical reaction

Atomic number – The number of protons in the nucleus of an atom

Avogadro's constant – 6×10^{23} (the number of particles in one mole)

Bioleaching – An extraction method that uses bacteria to extract metals from low-grade ores

Glossary

Blast furnace – Industrial method of extracting iron from iron ore

Carbon footprint – The total amount of carbon dioxide (and other greenhouse gases) emitted

Catalyst – A substance that changes the rate of a chemical reaction without being used up or chemically changed at the end of the reaction

Cathode – The negative electrode

Cation – A positive ion

Chromatography – A method of separating mixtures of dyes

Compound – A substance consisting of two or more elements chemically combined together

Cracking – A process used to break up large hydrocarbon molecules into smaller, more useful molecules.

Crystallisation – A method used to separate a soluble solid from its solution when you want to collect the solid

Delocalised electrons – Free-moving electrons

Displacement reaction – A reaction in which a more reactive element takes the place of a less reactive element in a compound

Electrolyte – A liquid or solution containing ions that is broken down during electrolysis

Element – A substance that consists of only one type of atom

Empirical formula – The simplest whole number ratio of each kind of atom present in a compound

Endothermic – A reaction in which energy is taken in

Energy level – A region in an atom where electrons are found

Equilibrium – A reversible reaction where the rate of the forward reaction is the same as the rate of the reverse reaction

Exothermic – A reaction in which energy is given out

Fertiliser – Any material added to the soil or applied to a plant to improve the supply of minerals and increase crop yield

Formulation – A mixture that has been designed as a useful product

Fossil fuel – Fuel formed in the ground over millions of years from the remains of dead plants and animal

Fractional distillation – A method used to separate mixtures of liquids

Fullerene – A molecule made of carbon atoms arranged as a hollow sphere

Halogen – One of the five non-metals in group 7 of the periodic table

High tensile strength – Does not break easily when stretched

Hydrocarbon – A molecule containing hydrogen and carbon atoms only

Intermolecular forces – The weak forces of attraction that occur between molecules

Ion – An atom or group of atoms that has gained or lost one or more electrons in order to gain a full outer shell

Isotopes – Atoms of the same element that have the same number of protons but different numbers of neutrons

Life-cycle assessment – An evaluation of the environmental impact of a product over the whole of its lifespan

HT Low-grade ores – Ores that contain small amounts of metal

Mass number – The total number of protons and neutrons in an atom

Mixture – Two or more elements or compounds that are not chemically combined

Mole – The amount of material containing 6×10^{23} particles

Monomer – The individual molecules that join together to form a polymer

Nanometre – 1×10^{-9} (0.000 000 001) m

Nanotube – A molecule made of carbon atoms arranged in a tubular structure

HT Ore – A naturally occurring mineral from which it is economically viable to extract a metal

HT Oxidation – A reaction involving the gain of oxygen or the loss of electrons

Particulates – Small solid particles present in the air

Peer-reviewed evidence – Work (evidence) of scientists that has been checked by other scientists to ensure that it is accurate and scientifically valid

Photosynthesis – The process by which green plants and algae use water and carbon dioxide to make glucose and oxygen

🔴 **Phytomining** – A method of metal extraction that involves growing plants in metal solutions so that they accumulate metal; the plants are then burnt and the metal extracted from the ash

Polymer – A large, long-chained molecule

Potable water – Water that is safe to drink

Precipitate – A solid formed when two solutions react together

Rate – A measure of the speed of a chemical reaction

🔴 **Redox** – A reaction in which both oxidation and reduction occur

🔴 **Reduction** – A reaction involving the loss of oxygen or the gain of electrons

Relative formula mass – The sum of the atomic masses of the atoms in a formula

Salt – A product of the reaction that occurs when an acid is neutralised

Shell – Another word for an energy level

Solute – A solid that dissolves in a liquid to form a solution

🔴 **Strong acid** – An acid that fully ionises when dissolved

Sustainable development – Living in a way that meets the needs of the current generation without compromising the potential of future generations to meet their own needs

Thermal decomposition – The breakdown of a chemical substance due to the action of heat

Unsaturated – A molecule that contains a carbon-carbon double bond

🔴 **Weak acid** – An acid that partially ionises when dissolved in water

Physics

ac – Alternating current that changes direction

Acceleration – Change in velocity over time

Atomic number – Number of protons in an atom

Becquerel – Unit of rate of radioactive decay

Braking distance – Distance a vehicle travels after the brakes have been applied

Compression – A region in a longitudinal wave where the particles are closer together

dc – Direct current that always passes in the same direction

Displacement – Distance travelled in a given direction

Elastically deformed – Stretched object that returns to its original length after the force is removed

Electric current – Flow of electrical charge

Emission – Release of energy

Gas pressure – Total force exerted by all of the gas molecules inside the container on a unit area of the wall

Gravity – Process where all objects with mass attract each other

Inelastically deformed – Stretched object that does not return to its original length after the force is removed

🔴 **Inertia** – The tendency of objects to continue in their state of rest or uniform motion

Irradiation – Exposing an object to nuclear radiation without the object becoming radioactive itself

Magnetic field – Region around a magnet where a force acts on another magnet or on a magnetic material

Mass number – Total number of protons and neutrons in an atom

Moment – Turning effect of a force

Momentum – Product of an object's mass and velocity

National Grid – System of transformers and cables linking power stations to consumers

Non-renewable – Energy resource that will eventually run out

Ohmic conductor – Resistor at constant temperature where the current is directly proportional to the potential difference

Parallel circuit – Circuit which contains branches and where all the components will have the same voltage

Permanent magnet – Magnet which produces its own magnetic field

Glossary

Radioactive decay – Random release of radiation from an unstable nucleus as it becomes more stable

Rarefaction – A region in a longitudinal wave where the particles are further apart

Reflection – Waves striking a boundary between different media and being returned back to the same media

Refraction – Change in direction of a wave as it travels from one medium to another

Renewable – Energy resource that can be replenished as it is used

Satellite – Object in orbit around a body

Scalar – A quantity that only has a magnitude

Series circuit – Circuit where all components are connected along a single path

Sievert – Unit of radiation dose

Solenoid – Coil wound into a helix shape

Specific heat capacity – Amount of energy required to raise the temperature of one kilogram of a substance by one degree Celsius

Specific latent heat – The energy required to change the state of one kilogram of the substance with no change in temperature

Thermal insulation – Material used to reduce transfer of heat energy

Thinking distance – Distance travelled before the driver applies the brakes

⊞ Transformer – Device used to increase or lower the voltage of an alternating current

Vector – A quantity that has both a magnitude and a direction

Velocity – Speed in a given direction

Weight – Force acting on an object due to gravity

The Periodic Table

Key

- Metals
- Non-metals

Key to element box:
- Relative atomic mass → 1
- Atomic symbol → H
- Name → hydrogen
- Atomic/proton number → 1

1	2											3	4	5	6	7	0 or 8
																	4 **He** helium 2
7 **Li** lithium 3	9 **Be** beryllium 4											11 **B** boron 5	12 **C** carbon 6	14 **N** nitrogen 7	16 **O** oxygen 8	19 **F** fluorine 9	20 **Ne** neon 10
23 **Na** sodium 11	24 **Mg** magnesium 12											27 **Al** aluminium 13	28 **Si** silicon 14	31 **P** phosphorus 15	32 **S** sulfur 16	35.5 **Cl** chlorine 17	40 **Ar** argon 18
39 **K** potassium 19	40 **Ca** calcium 20	45 **Sc** scandium 21	48 **Ti** titanium 22	51 **V** vanadium 23	52 **Cr** chromium 24	55 **Mn** manganese 25	56 **Fe** iron 26	59 **Co** cobalt 27	59 **Ni** nickel 28	63.5 **Cu** copper 29	65 **Zn** zinc 30	70 **Ga** gallium 31	73 **Ge** germanium 32	75 **As** arsenic 33	79 **Se** selenium 34	80 **Br** bromine 35	84 **Kr** krypton 36
85 **Rb** rubidium 37	88 **Sr** strontium 38	89 **Y** yttrium 39	91 **Zr** zirconium 40	93 **Nb** niobium 41	96 **Mo** molybdenum 42	[98] **Tc** technetium 43	101 **Ru** ruthenium 44	103 **Rh** rhodium 45	106 **Pd** palladium 46	108 **Ag** silver 47	112 **Cd** cadmium 48	115 **In** indium 49	119 **Sn** tin 50	122 **Sb** antimony 51	128 **Te** tellurium 52	127 **I** iodine 53	131 **Xe** xenon 54
133 **Cs** caesium 55	137 **Ba** barium 56	139 **La*** lanthanum 57	178 **Hf** hafnium 72	181 **Ta** tantalum 73	184 **W** tungsten 74	186 **Re** rhenium 75	190 **Os** osmium 76	192 **Ir** iridium 77	195 **Pt** platinum 78	197 **Au** gold 79	201 **Hg** mercury 80	204 **Tl** thallium 81	207 **Pb** lead 82	209 **Bi** bismuth 83	[209] **Po** polonium 84	[210] **At** astatine 85	[222] **Rn** radon 86
[223] **Fr** francium 87	[226] **Ra** radium 88	[227] **Ac*** actinium 89	[261] **Rf** rutherfordium 104	[262] **Db** dubnium 105	[266] **Sg** seaborgium 106	[264] **Bh** bohrium 107	[277] **Hs** hassium 108	[268] **Mt** meitnerium 109	[271] **Ds** darmstadtium 110	[272] **Rg** roentgenium 111	[285] **Cn** copernicium 112	[286] **Uut** ununtrium 113	[289] **Fl** flerovium 114	[289] **Uup** ununpentium 115	[293] **Lv** livermorium 116	[294] **Uus** ununseptium 117	[294] **Uuo** ununoctium 118

*The lanthanides (atomic numbers 58–71) and the actinides (atomic numbers 90–103) have been omitted.
The relative atomic masses of copper and chlorine have not been rounded to the nearest whole number.

Physics Equations

force = mass × acceleration

kinetic energy = 0.5 × mass × (speed)^2

work done = force × distance (along the line of action of the force)

$$\text{power} = \frac{\text{work done}}{\text{time}}$$

$$\text{efficiency} = \frac{\text{useful output energy transfer}}{\text{total input energy transfer}}$$

weight = mass × gravitational field strength *(g)*

gravitational potential energy = mass × gravitational field strength *(g)* × height

force applied to a spring = spring constant × extension

distance travelled = speed × time

$$\text{acceleration} = \frac{\text{change in velocity}}{\text{time taken}}$$

wave speed = frequency × wavelength

charge flow = current × time

potential difference = current × resistance

power = potential difference × current

power = (current)^2 × resistance

energy transferred = power × time

energy transferred = charge flow × potential difference

$$\text{density} = \frac{\text{mass}}{\text{volume}}$$

$$(\text{final velocity})^2 - (\text{initial velocity})^2 = 2 \times \text{acceleration} \times \text{distance}$$

change in thermal energy = mass × specific heat capacity × temperature change

thermal energy for a change of state = mass × specific latent heat

$$\text{elastic potential energy} = 0.5 \times \text{spring constant} \times (\text{extension})^2$$

$$\text{efficiency} = \frac{\text{useful power output}}{\text{total power input}}$$

$$\text{period} = \frac{1}{\text{frequency}}$$

HT

potential difference across primary coil × current in primary coil = potential difference across secondary coil × current in secondary coil

force on a conductor (at right angles to a magnetic field) carrying a current = magnetic flux density × current × length

momentum = mass × velocity

Index

Acknowledgements

The authors and publisher are grateful to the copyright holders for permission to use quoted materials and images.

All images are © Shutterstock and © HarperCollins*Publishers*.

Every effort has been made to trace copyright holders and obtain their permission for the use of copyright material. The authors and publisher will gladly receive information enabling them to rectify any error or omission in subsequent editions. All facts are correct at time of going to press.

Published by Collins
An imprint of HarperCollins*Publishers*
1 London Bridge Street
London SE1 9GF

HarperCollins*Publishers*
1st Floor, Watermarque Building,
Ringsend Road, Dublin 4, Ireland

ISBN: 9780008276072

First published 2018
This edition published in 2020

Previously published as Letts

10 9 8 7 6 5

© HarperCollins*Publishers* Limited 2020

All rights reserved. No part of this publication may be reproduced, stored in a retrieval system, or transmitted, in any form or by any means, electronic, mechanical, photocopying, recording or otherwise, without the prior permission of Collins.

British Library Cataloguing in Publication Data.

A CIP record of this book is available from the British Library.

Authors: Tom Adams, Dan Evans and Dan Foulder
Commissioning Editors: Clare Souza and Kerry Ferguson
Editor/Project Manager: Katie Galloway
Cover Design: Kevin Robbins and Gordon MacGilp
Inside Concept Design: Ian Wrigley
Text Design and Layout: Nicola Lancashire at Rose & Thorn Creative Services, and Ian Wrigley
Production: Natalia Rebow
Printed and bound in the UK using 100% Renewable Electricity at CPI Group (UK) Ltd

MIX
Paper from
responsible source
FSC www.fsc.org   **FSC® C007454**

This book is produced from independently certified FSC™ paper to ensure responsible forest management.

For more information visit:
www.harpercollins.co.uk/green